T0182091

Harvesting Rainwater from Buildings

Syed Azizul Haq, PEng

Harvesting Rainwater from Buildings

 Springer

Syed Azizul Haq, PEng
Public Works Department
Dhaka
Bangladesh

ISBN 978-3-319-83500-6   ISBN 978-3-319-46362-9   (eBook)
DOI 10.1007/978-3-319-46362-9

Printed on acid-free paper

This Springer imprint is published by Springer Nature
The registered company is Springer International Publishing AG
The registered company address is: Gewerbestrasse 11, 6330 Cham, Switzerland

To my wife
Dr. Farida Hashmat Jahan

Foreword I

I would offer my congratulations on the publishing of *Harvesting Rainwater from Buildings*. I am honored to be writing the foreword and, in my capacity as International Advisor of the Bangladesh Rainwater Forum Society and president of Group Rain Drops—which published *Rainwater & You*—I am very happy that this vitally important concept of rainwater harvesting will be further disseminated in Bangladesh through the publication of this book.

From "drain city to rain city," from "dependence on off-site water resources to independence on on-site water resources," and from "life-line to life points"—This is the "Tokyo lesson." Sumida Ward in Tokyo suffered from severe urban flooding as well as water shortages and major earthquakes in the 20th century. We started rainwater-harvesting projects in 1982 for flood control, saving the city water supply, securing emergency water, and regenerating urban-water cycles. I have been involved in rainwater-harvesting projects as chief of the Rainwater Harvesting Promoting section in the Sumida municipal government. Sumida took initiatives to promote rainwater harvesting at a public level as well as private and community levels, in cooperation with People for Rainwater, an NPO established in 1995, after the Tokyo International Rainwater Utilization Conference was held in Sumida city in 1994.

The Sumida Rainwater Rules were established in 2008 to promote urban development. Based on these regulations, Tokyo Sky Tree, which is the highest broadcasting tower in the world and was completed in 2012, has become a symbol of climate change adaptation in Tokyo. Rainwater falling on the roof of the observation platforms at 350 and 450 m as well as the adjacent buildings has been stocked into a basement tank with a total capacity of 2635 tons. This means that even if the peak rainfall of 100 mm falls on Tokyo, enough rainwater will flow into the rain tanks to prevent flooding.

Harvested rain has been used for watering the green roof and for emergency water. Rainwater-harvesting systems have been introduced into 560 private houses and buildings in Sumida City, and the total capacity of rainwater tanks has reached 22,000 tons. Sumida City has moved from a "rain city" to a "drain city."

Following the initiatives of Sumida City, 208 local governments set up subsidy systems for rainwater utilization and infiltration all around Japan. In addition, the Japan Rainwater Harvesting Promotion law was established in 2014. According to this law, a national guideline for rainwater harvesting was set up in 2015. It states that rainwater-harvesting systems should, in principle, be introduced into new national public buildings. In addition, the Japanese central government has recommended local governments to promote rainwater harvesting all around Japan.

To try and integrate sustainable rainwater harvesting into the fabric of society, in 1995 I published (in Japanese) *Rainwater & You: 100 Ways to Use Rainwater.*" which covers background, policy, design, and case studies of rainwater harvesting not only in Japan but also all around the world. The book was translated into nine languages and published in eight countries. The Architectural Institute of Japan published *Technological Standards on Rainwater Harvesting* in March 2016. It will be useful for architects and plumbers who design and install rainwater-harvesting systems in buildings and houses. Industrial national-standard technologies and products of rainwater-harvesting systems, such as those established in Germany (as DIN [German Industrial Standard]) should be considered in order to secure low-cost, high-quality, long-life and easy-maintenance systems in the future.

After the mainstreaming of Japanese rainwater harvesting in urban areas, it was good timing to publish *Harvesting Rainwater from Buildings* in Bangladesh. It covers basic information about rainwater harvesting such as water requirements, rainfall, roof, tank storage, water treatment, water supply, ground recharging, and drainage. Mr. Haq, the author of this book, is additional chief engineer of the Public Works Department in Bangladesh as well as president of the Bangladesh Rain Forum society. He is a specialist of plumbing in the field of civil engineering, and in this book he proposes special designs of rainwater storage for buildings. This book would be most helpful for architects and plumbers who design and install rainwater harvesting systems in buildings not only in Bangladesh but also those who do so internationally. I hope that the Bangladesh government encourages introducing rainwater-harvesting systems into new buildings to solve the water crisis in mega cities such as Dhaka city, Chittagong city, and Khulna city. I'm sure that *Harvesting Rainwater from Buildings* can make a great contribution to develop technological national standards in Bangladesh.

Japan and Bangladesh are blessed with rain and beneath the same sky of Monsoon Asia. Both countries have suffered from urban flood, cyclone, and water shortages. The problem is insufficient water; the solution is rainwater harvesting! I sincerely hope we can learn water lessons from each other and share and exchange experiences of rainwater harvesting. I believe this book is a big step toward that goal.

Shizuoka, Japan   Dr. Makoto Murase, Ph.D.
April 2016   Visiting Professor of Faculty of
                                                              Pharmaceutical Sciences
                                                              Toho University, Japan
                                        Chairman, Skywater Bangladesh Ltd.

Foreword II

I read the manuscript on Harvesting Rainwater from Buildings written by Syed Azizul Haq, PEng, with keen interest. Earlier in my life I taught hydrology for a number of years to the students of Civil Engineering at the Bangladesh University of Engineering and Technology (BUET) where I was a member of the faculty. Therefore, issues related to the characteristics of rainfall, the relationship between rainfall and runoff, issues related to groundwater recharge, etc., are of interest to me. I read the chapters on groundwater recharge, rainwater drainage, and rainwater-harvesting elements with due attention. The book introduces these issues in a brief but succinct manner; therefore, any reader will be able to develop a clear understanding of the hydrological cycle and the availability of rainwater that can be harvested in an economical manner for use in a building for various purposes.

Bangladesh receives very high annual but limited rainfall over a period of 3–4 months each year. The average annual rainfall in Bangladesh is more than 2500 mm. It varies from 1500 mm on the western side to 10,000 mm in the northeast. However, because the rainfall is limited over the months of June through September, it has never been considered as a regular and dependable source of supply for drinking water throughout the year.

Reports are available of people of the coastal belt who, in earlier days, used to store rainwater in large earthen jars for consumption throughout the year. However, thousand of ponds and tanks were excavated all over the country to store rainwater to be used as drinking water over a period of one year. Harvesting rainwater from buildings in urban areas is an emerging concept in our country. The government is showing keen interest in adapting rainwater harvesting as an important source of water in commercial as well as residential buildings. The book very clearly describes possible options for the collection of rainwater as well as its storage and subsequent use for various domestic purposes including sanitation. The chapters on estimating the demand for water in various types of buildings—as well as how to collect, store, and process rainwater—are important sections of the book.

In my view, any professional—as well as anyone interested in harvesting rainwater—will find the book to be a practical and professional guide. The book demonstrates that in a world where water is a finite resource, it is an economic as well as environmental and social good; however, rainwater is a free good that must be harvested, processed, and used appropriately. Water is one of the essential elements for sustaining life and livelihood in the world. With some effort, rainwater that occurs naturally can be a valuable resource that can be used for benefit for humankind and society.

I am confident that the book will gain popularity among all those interested in harvesting rainwater from buildings.

Dr. Ainun Nishat
Professor Emeritus, Centre for Environmental and Climate
Change Research, Ex-Vice Chancellor, BRAC University,
Bangladesh

Preface

In the year 2000, the Ministry of Housing and Public Works (MoHPW) instructed the Public Works Department (PWD) to write a guideline on rainwater harvesting. Finally a booklet on rainwater harvesting was published jointly by the Public Works Department and the Department of Architecture (DoA) under the same ministry. The booklet was merely an introduction to rainwater harvesting furnishing some guidelines that were applicable to small-dwelling houses. Limited knowledge and availability of information in this regard were the basic reason behind the publishing of such a booklet of rainwater-harvesting guidelines.

Engineer Quamrul Islam Sidique, PEng, at that time the secretary of MoHPW, arranged the visit of Dr. Makoto Murase of Japan, nicknamed "Dr. Rain" and now known as "Dr. Skywater" (and who wrote one of the forwards for this book) came to Bangladesh to share his long-standing and vast experience in this area with engineers and architects in the departments under the ministry. Subsequently, a few piloting initiatives of harvesting rainwater from buildings, mostly projects, were implemented at a small scale.

With time, the overall water situation, particularly the water supply for cities and urban centers, in terms of both quality and quantity, was increasingly believed to be declining considerably. Until today, among water users, buildings of various occupancies are the major consumers and worse sufferers of the water crisis. In this situation, although rainwater harvesting was thought to be a reliable mitigating option, little heed was taken among policy makers, building regulators, and developers—in both the public and private sectors—to take pragmatic steps in this perspective. In these efforts, various important issues surfaced regarding the implementation of this concept predominantly in building sectors. The major issues were the lack of exclusive guidelines covering all of the aspects of rainwater harvesting suitable for different local conditions plus ignorance regarding plumbing, which plays a vital role in rainwater harvesting, particularly in buildings.

Against this backdrop, with a view toward disseminating the concept of rainwater harvesting, the underwriter introduced rainwater management as an exclusive chapter in his first endeavor, *Plumbing Practices* in 2006, which appeared simply as an introduction to harvesting rainwater from buildings. However, to obtain exclusive knowledge on harvesting rainwater from buildings, a blend of knowledge about plumbing and rainwater harvesting would be needed, which is considered the theme of the book.

This book elaborately describes all aspects of rainwater harvesting—including the basic concepts, procedures, opportunities, and practice of rainwater harvesting —mainly focusing its application in buildings of various occupancies and sizes. The text of the book is written in such a way as to provide a user-friendly methodology for the planning, design, construction, and maintenance of rainwater-harvesting infrastructure in buildings and their premises as a supplement to conventional public or private piped-water supplies. Eleven chapters are incorporated into the book. Very brief introductory notes of all the chapters are as follows.

Chapter 1—*Introduction*: An introduction to rainwater harvesting and buildings is provided. Ways of using rainwater in buildings directly or indirectly to meet the total demand for water, either partially or fully, are discussed.

Chapter 2—*Rain and Rainwater*: To understand rainwater harvesting, knowledge regarding rain, rainfall, and rainwater is a must. As such, the characteristics and properties of rain, rainfall intensity, and rainwater quality are discussed in this chapter.

Chapter 3—*Rainwater-Harvesting Technologies*: The development of a rainwater-harvesting system in a building is a technology-based work. Technological complicacy arises with the complicacy of a building configuration, in which the planning of a harvesting system emerges as another important concern. Therefore, various technical issues and planning approaches to harvesting rainwater from buildings are discussed exclusively herein.

Chapter 4—*Water Requirements*: To harvest rainwater in a building, it is first required to determine the amount of rainwater to be harvested. For this purpose, the total water requirement for the building should be estimated judiciously. As such, the various water requirements in a building of various occupancies should be known. In this chapter, water requirements for various purposes in buildings of different occupancies are discussed and tabulated.

Chapter 5—*Rainwater Collection*: Rainwater collection is the first job of rainwater harvesting. Approaches to rainwater collection differ with varieties in the configuration, size, and location of catchments. Various collection techniques of rainwater falling on various natures of exposed surfaces of buildings and ways of maximizing the collection of rainwater are discussed in this chapter.

Chapter 6—*Rainwater Storage*: Storing rainwater is the most important factor in calculating the cost of the system to be developed. The second important aspect is managing sufficient space for storage inside or outside the building. All concerns regarding the storage of rainwater in buildings are delineated in this chapter. The merits and demerits of storing rainwater at various locations in buildings are also discussed.

Chapter 7—*Rainwater Conditioning*: The conditioning of rainwater for its harvest is an optional job depending on the quality of rainwater subject to its purpose of use. Various aspects of rainwater conditioning are discussed in this chapter. Boiling, chlorination, UV and ozonation, screening, sedimentation, and filtration processes are discussed as is the well-practiced conditioning system suitable for harvesting rainwater from buildings.

Chapter 8—*Rainwater-supply System*: In buildings, rainwater may be needed at various locations, in which case it must be supplied from a rainwater-storage tank. Various types of rainwater-distribution systems, independently or in conjunction with a normal water-supply system, are discussed. The planning and designing of various components of such a supply system are also described in this chapter.

Chapter 9—*Groundwater Recharging*: Groundwater recharging using rainwater is another aspect of rainwater harvesting. Groundwater recharging might be an obligatory function to be performed by the building owner or developer. As such, all methods of recharging—including the injection well and the measures to be taken in this regard—are addressed in this chapter.

Chapter 10—*Rainwater Drainage*: In the case of heavy rainfall, or when a small amount of rainwater is to be harvested, excess rainwater must be drained out properly. Various ways of draining excess rainwater from buildings and their premises are discussed in this chapter.

Chapter 11—*Rainwater-harvesting Elements*: Various elements must be incorporated in developing a rainwater-harvesting system in a building. Knowledge regarding those elements will help in developing a sustainable system. Important aspects of the elements used in rainwater harvesting in buildings are discussed very briefly in this chapter.

In summary, this book would be highly beneficial for professionals, engineers, architects, and others concerned—including students and policy makers—requiring guidance or information regarding rain and rainwater harvesting, particularly in buildings of all sizes and occupancies.

Dhaka, Bangladesh   Syed Azizul Haq, PEng

Acknowledgements

At the onset, I wholeheartedly acknowledge the blessing of Almighty Allah, the most gracious and most merciful, Who blessed me with all sorts of knowledge and information needed to accomplish this book.

Next, I humbly acknowledge the continuous encouragement of my mother, who used to pray to Almighty Allah to grant me the ability to write this book.

I owe my profound gratitude to Dr. Afzal Ahmed and Dr. Tanvir Ahmed. Without their heartfelt support, by going through the manuscript and making valuable suggestions and comments, it would not have been possible to shape my ideas and enrich the book.

My special thanks and gratitude goes to my mentor, Dr. Feroz Ahmed, for his valuable insights, encouragement, and support during preparation of the manuscript. I am also indebted to Dr. Makoto Murase and Dr. Inun Nishat, both of whom enlightened the book by furnishing their invaluable forewords.

I am also indebted to Engr. Abdul Malek Sikder, Engr. Qumruzzaman, and Engr. Md. Riadur Rahman of PWD, who helped me a lot in improving the write-up and furnishing the finished work.

Last, although I regret that I cannot mention all of their names individually, I am very grateful to everyone who encouraged me with complementary words and helped me in a diversified manner during the long journey toward publication of this book.

Syed Azizul Haq, PEng

Contents

1   Introduction 1
  1.1   Introduction 2
  1.2   Surface Water 2
  1.3   Groundwater 3
  1.4   Rainwater Utilization 5
  1.5   History of Rainwater Harvesting 6
  1.6   Rainwater Harvesting Around the World 7
       1.6.1   Rainwater Harvesting in Bangladesh 8
       1.6.2   Rainwater Harvesting in India 8
       1.6.3   Rainwater Harvesting in China 9
       1.6.4   Rainwater Harvesting in Africa 9
       1.6.5   Rainwater Harvesting in Japan 10
       1.6.6   Rainwater Harvesting in Germany 10
       1.6.7   Rainwater Harvesting in the United Kingdom 10
       1.6.8   Rainwater Harvesting in Australia 11
       1.6.9   Rainwater Harvesting in the United States
             of America 11
  1.7   Building 12
  1.8   Rainwater Use in Buildings 13
  1.9   Establishing a Rainwater-Harvesting System 14
       1.9.1   Formulating Policy 15
       1.9.2   Creating Awareness 15
       1.9.3   Instituting Incentives 15
       1.9.4   Enforcing Rules and Regulations 16
       1.9.5   Formulating Codes and Guidelines 16
       1.9.6   Building Capacity to Act 17
  1.10   Challenges to Rainwater Harvesting in Buildings 17
       1.10.1 Changing Perceptions 17
       1.10.2 Conflicting and Missing Guidelines in Codes 18

	1.10.3	Standardizing Water Quality and Water-Testing Protocols	18
	1.10.4	Absence of Good Governance	19
References			19

2 Rain and Rainwater 23
2.1 Introduction 23
2.2 Clouds 24
2.3 Rain 24
2.4 Rainfall 25
    2.4.1 Amount of Rainfall 25
    2.4.2 Rainfall Intensity 25
    2.4.3 Rainfall Distribution 27
    2.4.4 Rainfall Measurement 27
    2.4.5 Rain Gauges 28
    2.4.6 Ordinary Rain Gauge 28
    2.4.7 Installing the Rain Gauge 29
    2.4.8 Rain Measurement 29
    2.4.9 Observations 30
    2.4.10 Maintenance 31
2.5 Global Rainfall Scenario 31
    2.5.1 Africa 31
    2.5.2 North America 32
    2.5.3 South America 32
    2.5.4 Antarctica 33
    2.5.5 Australia 33
    2.5.6 Europe 33
    2.5.7 Asia 34
2.6 Rainfall in India 35
2.7 Rainfall in Bangladesh 36
2.8 Rainwater Quality 38
    2.8.1 Pollutants in Rainwater 38
    2.8.2 Microorganisms in Rainwater 39
    2.8.3 Chemical Contamination of Rainwater 39
    2.8.4 Qualitative Changes in Rainwater 39
References 41

3 Rainwater-Harvesting Technology 45
3.1 Introduction 45
3.2 Scopes of Rainwater Harvesting 46
    3.2.1 Rainwater for General-Purposes Use 46
    3.2.2 Rainwater for Groundwater Recharging 46

3.3 Functional Techniques in Rainwater Harvesting 46
    3.3.1 Functional Techniques for General-Use 47
    3.3.2 Functional Techniques for Groundwater
           Recharge 47
3.4 Extent of Technological Involvement 47
    3.4.1 Direct-Use System 48
    3.4.2 Nonfiltered System 49
    3.4.3 Filtered System 50
    3.4.4 Complete System 50
3.5 Aspects of Rainwater-Harvesting Technology 51
    3.5.1 Planning Aspects of a Rainwater-Harvesting
           System 51
    3.5.2 Design Aspects of a Rainwater-Harvesting
           System 51
    3.5.3 Construction Aspects of a Rainwater-Harvesting
           System 52
    3.5.4 Maintenance Aspects of a Rainwater-Harvesting
           System 52
3.6 Planning Approach for a Rainwater-Harvesting System 53
    3.6.1 Identifying the Purpose of Use 53
    3.6.2 Planning the Catchments 53
    3.6.3 Planning the Collection 53
    3.6.4 Planning the Storage System 54
    3.6.5 Planning the Conditioning System 54
3.7 Design Approach of a Rainwater-Harvesting System 54
    3.7.1 Estimating the Amount of Rainwater 54
    3.7.2 Sizing the Catchment Area 54
    3.7.3 Designing of Collection and Overflow Pipes 55
    3.7.4 Designing Other Components 55
3.8 Construction Approach for a Rainwater-Harvesting System . . . 55
3.9 Maintenance Approach to a Rainwater-Harvesting System 56
3.10 Prospects of Rainwater Harvesting 57
3.11 Problems of a Rainwater-Harvesting System 57
References 58

4 Water Requirement 59
4.1 Introduction 59
4.2 Various Types of Water Demand 60
4.3 Domestic Water Demand 61
4.4 Institutional and Commercial Water Demand 62
    4.4.1 Water Use in Office Buildings 62
4.5 Water Demand for Common Utility Purposes 62
4.6 Water Demand for Other Purposes 63
    4.6.1 Water for Fire Fighting 64
    4.6.2 Recreational Water 65

	4.6.3	Water for Gardening and Plantation............	66
	4.6.4	Water for Animal Rearing.....................	66
	4.6.5	Water for Special Uses	67
4.7		Loss and Wastage of Water	68
4.8		Factors Affecting Per-Capita Demand..................	69
	4.8.1	Climatic Conditions..........................	69
	4.8.2	Size of Community	69
	4.8.3	Living Standard of the People.................	69
	4.8.4	Manufacturing and Commercial Activities........	70
	4.8.5	Pressure in the Distribution System.............	70
	4.8.6	Sanitation and Drainage System	70
	4.8.7	Cost of Water	71
	References..		71

5	**Rainwater Collection**.......................................		73
5.1		Introduction	73
5.2		Catchments..	74
	5.2.1	Building Elements as Catchments	74
	5.2.2	Planning the Catchments	74
5.3		Rainwater Collection from a Roof	75
5.4		Types of Roofs.......................................	76
	5.4.1	Flat Roof.................................	76
	5.4.2	Sloped Roofs..............................	77
	5.4.3	Folded Roofs..............................	79
5.5		Catchment-Surface Materials...........................	80
	5.5.1	Metal Surface	81
	5.5.2	Clay and Concrete Tile	82
	5.5.3	Composite or Asphalt Shingle.................	82
	5.5.4	Other Roofing Materials	82
5.6		Rainwater Collection from a Roof	83
	5.6.1	Collection from a Flat Roof....................	83
	5.6.2	Collection from a Sloped or Curved Roof.........	84
	5.6.3	Collection from a Folded Roof	85
5.7		Effective Catchment Area.............................	85
	5.7.1	Catchments of a Flat-Roof Surface	86
	5.7.2	Catchment of an Inclined Surface	87
5.8		Rainwater-Conveying Media...........................	88
	5.8.1	Gutter	88
	5.8.2	Rainwater Down Pipe (RDP)...................	89
5.9		First-Flush Diversion	90
	5.9.1	Sizing the First Flush........................	91
	5.9.2	Methods of First-Flush Diversion	92
	5.9.3	Manual First-Flush System	92
	5.9.4	Floating-Ball First-Flush System	93

5.10 Techniques for Maximizing Rainwater Collection 94
      5.10.1 Planning the Collection Elements 94
      5.10.2 Increasing the Catchment Area 94
      5.10.3 Choosing the Finished Surface 95
      5.10.4 Choosing the Appropriate Collection Elements 95
      5.10.5 Designing the Collection Elements 96
      5.10.6 Good Workmanship. 96
References. 96

6 Rainwater Storage. 99
  6.1 Introduction 99
  6.2 Storing Rainwater. 100
      6.2.1 Factors of Storage Development 100
  6.3 Location of Storage 100
      6.3.1 Storage Inside the Building 101
      6.3.2 Storage Outside of the Building. 104
  6.4 Planning of Rainwater Storage 106
      6.4.1 Separate Storage and Supply System. 107
      6.4.2 Combined Storage and Supply System 108
  6.5 Sizing of the Storage Tank. 109
      6.5.1 Demand-Side Approach. 109
      6.5.2 Supply-Side Approach. 110
      6.5.3 Shape of Storage Tank 111
  6.6 Aesthetics. 112
  6.7 Functionality of a Storage Tank 113
  6.8 Prefabricated and Constructed Tanks 114
      6.8.1 Prefabricated Tanks. 114
      6.8.2 Constructed Tanks. 115
      6.8.3 Sectional Tank. 115
References. 115

7 Rainwater Conditioning 117
  7.1 Introduction 117
  7.2 Methods of Conditioning Rainwater. 118
  7.3 Screening 118
  7.4 Sedimentation. 120
      7.4.1 Design Aspects of Sedimentation Tanks 120
  7.5 Filtration. 122
      7.5.1 Gravity Filters 124
      7.5.2 Pressure Filters 125
  7.6 Disinfection 126
      7.6.1 Boiling. 127
      7.6.2 Ultraviolet Light 128
      7.6.3 Chlorination. 129
      7.6.4 Ozonation 131

7.7   Planning the Conditioning System 132
References... 132

8   Rainwater Supply System............................... 135
8.1   Introduction 135
8.2   Rainwater-Distribution Approach 135
         8.2.1   Exclusive Rainwater Supply 136
         8.2.2   Supplemental Water Supply................... 136
8.3   Rainwater-Distribution System 137
         8.3.1   Underground-Overhead Tank System 138
         8.3.2   Direct-Pumping System...................... 138
8.4   Pump and Pumping................................... 140
         8.4.1   Head of Pump.............................. 141
         8.4.2   Flow of Pump 143
         8.4.3   Frictional Loss in a Pipe 144
         8.4.4   Efficiency of a Pump........................ 146
         8.4.5   Power of a Pump........................... 146
8.5   Pipe Sizing 147
         8.5.1   Sizing Pipe for Suction and Delivery.......... 147
         8.5.2   Sizing of Pipe for Riser and Branch 148
8.6   General Requirement 151
References... 151

9   Groundwater Recharging............................... 153
9.1   Introduction 153
9.2   Groundwater Recharging 154
         9.2.1   Positive Impacts of Recharging 154
         9.2.2   Negative Impacts of Recharging 155
9.3   Information Needed 155
9.4   Recharge Structure–Design Parameters.................. 156
         9.4.1   Infiltration Rate 156
         9.4.2   Retention Time 157
         9.4.3   Effective Porosity.......................... 157
         9.4.4   Permeability................................ 158
         9.4.5   Hydraulic Transmissivity..................... 158
9.5   Methodology of Recharging 158
         9.5.1   Natural Recharging 159
         9.5.2   Artificial Recharging 161
References... 171

10   Rainwater Drainage 173
10.1   Introduction 173
10.2   Draining System...................................... 174
         10.2.1   Gravitational System 174
         10.2.2   Pumping System 174

	10.3	Gravitational-Drainage Design Factors	175
		10.3.1 Imperviousness of the Surface	175
		10.3.2 Time of Concentration	175
		10.3.3 Hourly Intensity of Rainfall	176
	10.4	Storm Water–Disposal Guidelines	177
	10.5	Surface Drains	178
		10.5.1 Computation of Peak Runoff	179
		10.5.2 Computation of Discharge Capacity	179
		10.5.3 Surface-Drain Design Considerations	180
		10.5.4 Drain-Configuration Considerations	182
	10.6	Subsurface Piped Drainage	183
		10.6.1 Sizing of Stormwater–Drainage Piping	183
		10.6.2 Structural Safety of Stormwater-Drainage Piping	184
	10.7	Grading of Land Surface	185
	10.8	Manholes or Inspection Pits	186
		10.8.1 Location	187
		10.8.2 Size of Inspection Pits or Manholes	187
	References		189
11	**Rainwater-Harvesting Elements**		**191**
	11.1	Introduction	191
	11.2	Elements in Rainwater Harvesting	191
	11.3	Pipe	192
		11.3.1 Galvanized Iron Pipe	192
		11.3.2 Plastic Pipe	193
		11.3.3 Cast Iron Pipe	194
		11.3.4 Concrete Pipe	195
	11.4	Pipe Fittings	196
		11.4.1 Fittings for Jointing	196
		11.4.2 Fittings for Changing Pipe Diameter	198
		11.4.3 Fittings for Changing Direction	199
		11.4.4 Fittings for Pipe Branching	199
		11.4.5 Fittings for Pipe Closing	201
	11.5	Pipe Jointing	202
		11.5.1 GI Pipe Jointing	202
		11.5.2 Plastic-Pipe Jointing	203
		11.5.3 Cast-Iron Pipe Jointing	204
		11.5.4 Concrete-Pipe Jointing	205
		11.5.5 Pipe-Joint Testing	205
	11.6	Supporting Pipes	206
	11.7	Disinfecting Piping	206
		11.7.1 Disinfecting and Cleaning the System	208

11.8   Storage Tanks...   208
        11.8.1   RCC or Masonry Tanks.......................   209
        11.8.2   Ferro-cement Tank...........................   209
        11.8.3   G.I. Tank....................................   210
        11.8.4   Stainless Steel Tank.........................   210
        11.8.5   Plastic Tank.................................   211
11.9   Pump   211
        11.9.1   Submersible Pumps   211
        11.9.2   Pump Installation............................   213
        11.9.3   Pump Room..................................   213
11.10   Valve   215
        11.10.1   On–off Valve................................   216
        11.10.2   Throttling Valve   216
        11.10.3   Check Valve   217
        11.10.4   Float Valves   218
        11.10.5   Pressure-Reducing Valve......................   218
        11.10.6   Air-Release Valve   218
        11.10.7   Flush Valve..................................   219
11.11   Cock..   219
11.12   Faucet...   221
        11.12.1   Compression Faucets.........................   221
        11.12.2   Fuller Faucet   223
        11.12.3   Mixture Faucets..............................   223
11.13   Water Meter   223
        11.13.1   Disc Meter...................................   224
        11.13.2   Turbine Meter   224
        11.13.3   Compound Meter.............................   225
11.14   Elements in Recharge Structures............................   225
        11.14.1   Bricks and Cement Blocks....................   225
        11.14.2   Cement   226
        11.14.3   Aggregate   227
        11.14.4   Sand   227
        11.14.5   Reinforcing Bar..............................   227
        11.14.6   Cement–Sand Mortar.........................   228
        11.14.7   Concrete   228
        11.14.8   Water.......................................   229
        11.14.9   Reinforced Cement Concrete..................   230
11.15   Filter Media   230
        11.15.1   Filtering Sand   231
        11.15.2   Filtering Gravel..............................   231

11.16   Drainage Structures................................ 231
         11.16.1   Drainage Pipe 232
         11.16.2   Inspection Pit or Manhole Construction........... 233
         11.16.3   Manhole Cover 234
    References.. 236

Appendix... 239

Index 263

Abbreviations

ABS	Acrylonitrile Butadiene Styrene
AD	Anno Domini
BC	Before Christ
BHP	Brake Horse Power
BWG	Birmingham Wire Gauge
CI	Cast Iron
cm^2	Square Centimeter
cm	Centimeter
CPVC	Chlorinated Polyvinyl Chloride
DN	Diameter Nominal
g	Galon
GI	Galvanized Iron
h	Hour
H	Head
kg	Kilogram
kN	KiloNewton
kPa	Kilo Pascle
lpcd	Liter per capita per day
lpm	Liter per minute
L	Litre
Lps	Liter per second
m^3	Cubic meter
m	Meter
mm	Millimetre
MPa	Mega Pascle
N	Newton
NTU	Nephelometric Turbidity Units
OH	Overhead
OPC	Ordinary Portland Cement
Pa	Pascle

pH	Potential of Hydrogen
PPC	Pozzolanic Portland Cement
ppm	Parts per million
PVC	Polyvinyl Chloride
RCC	Reinforced Cement Concrete
RDP	Rainwater Down Pipe
s	Second
spp	Species
3D	Three dimension
TSP	Trisodium phosphate
μm	Micro-meter
UG	Underground
uPVC	Unplastisized polyvinyl chloride
UV	Ultraviolet
VWLP	Vertical Water-Lifting Pipe
WC	Water Closet
WHO	World Health Organization
WSFU	Water Supply–Fixture Units
y	Year

Chapter 1
Introduction

Abstract Water is life and a primary need of all living beings. Human beings also need water not only for survival but also for their multipurpose and ever increasing needs other than the basics. The amount of usable water on the Earth's surface is continuously shrinking and becoming polluted due to various human activities in the name of development. When surface waters, as a primary choice, are being polluted daily, then underground water as a secondary option is continually being extracted in many parts of the globe while the negative impacts of this practice are ignored. In this context, a third option is investigated herein. Buildings of various occupancies are the only infrastructures in which human beings live or conduct various businesses safely and comfortably. For safe and comfortable living in buildings, even for a little while, there is need of water. Providing wholesome water in buildings is therefore a major concern in building development. For buildings, water is generally drawn from a public main running by the side of the building; in unavoidable cases, an independent water source, such as tube well, is used for withdrawing underground water. In many parts of the globe, both surface and groundwater is continually becoming limited due to natural calamities and man-made interventions; in contrast, the demand for building development is increasing with the increase in population and the quest for socioeconomic development. The development of buildings is therefore creating an increase in water demand, which cannot be met by the usual practices depending on the sources of conventional surface and groundwater. Against this backdrop, a third option is envisaged for managing the shortage of water for any building: The practice of rainwater harvesting has been found to be reliable where sufficient rainfall occurs and adequate catchment is available to collect rainwater. In this chapter, the circumstances invoking rainwater harvesting in buildings is discussed, and its history of practice—as well as the present scenario in the global perspective—are delineated. An introduction to buildings and the prospects and the constraints of using rainwater in buildings is also described.

© Springer International Publishing Switzerland 2017
S.A. Haq, PEng, *Harvesting Rainwater from Buildings*,
DOI 10.1007/978-3-319-46362-9_1

1.1 Introduction

Rainwater is the root source of all water, both surface water and groundwater, both of which are mostly used as primary sources of water supply. Heat from the Sun causes the water of seas, rivers, the Earth's surface, etc. to evaporate and rises upward. At high altitude, rising water vapour expands by absorbing energy from the surrounding air due to a reduction in the atmospheric pressure and thus cools down. When the atmospheric temperature falls below the dewpoint temperature, the moist air cannot retain the excessive moisture and starts falling in the form of rain, hail, dew, sleet, frost, or precipitation. This process of evaporation and precipitation continues in cyclic order.

Among all of the types of precipitation, only rain will be the main focus here. The discussions will mainly cover when, where, and how this rainwater can be used in buildings, directly or indirectly, to meet the demand for total water either partially or fully.

1.2 Surface Water

When rain falls on the ground, directly or indirectly, a major portion of it generally flows over the ground's surface. The term "surface water" refers to those waters on the Earth's surface that flow in streams and rivers as well as water in artificial or man-made lake, canals, ponds, etc. Seawater, which is salty, also falls into the group of surface waters.

Among the surface waters, river water, as a source of sweet water, has been found to be the best choice. This quality of primary and potential source of water from nature is believed to be the main reason behind ancient civilizations developing at the banks of many rivers on the globe. Dhaka, the capital city of Bangladesh, was also started by the bank of the River Buriganga due to the river's sweet water. With the continuous sprawling of urbanization and the growth of various industries, surface waters become polluted due to the discharge of treated, partially treated, or even untreated wastes. When the pollution potential of the surface waters exceeds the manageable limit, dependency on groundwater sources, as the second alternative option, increases.

In underdeveloped countries, most of the urban centers or cities, which are usually situated near or by the bank of a river, discharge their treated, partially treated, and even untreated wastewater and sewage into the rivers. Moreover, industries tend to be increasingly built on river banks, or near rivers or canals, and discharge their treated or untreated wastes into those water bodies, thus causing heavy pollution. There are areas where pollution levels in river water during dry periods are found to be comparatively high due to having less water flow. In these circumstances, the treatment cost of water for supply increases and may exceed

Fig. 1.1 Pollution in surface water of the River Buriganga of Dhaka, Bangladesh [1] (*source* http://bdnature1.blogspot.com/2009/05/water-pollution-in-buriganga.html)

beyond the cost-effective limit. Figure 1.1 shows a highly polluted river in Bangladesh.

Again, for areas far away from any potential surface-water sources, the cost for conveying water may be too high to be cost-effective. In these circumstances, underground water, if available, is considered to be the next alternative source of water.

1.3   Groundwater

Groundwater is an important and potential source of water for personal and commercial uses. The Earth's surface comprises different layers of soil mass such as sand, gravel, clay, rock, etc. The layers of rock or compact clay, being solid or nonporous, cannot store water. In contrast, the layers of coarse sand and gravel have many pores and cracks, which allow rainfall to enter into the soil and start percolating from the natural surface. The porous soil layers, which remain filled with water, are called "aquifers." In most cases, groundwater has a tendency to flow very slowly toward lower levels. In highly undulating terrain, particularly hilly areas, the aquifer might become connected with the ground surface where the groundwater flows out of the soil into, e.g., a river or a spring. The surface waters evaporate

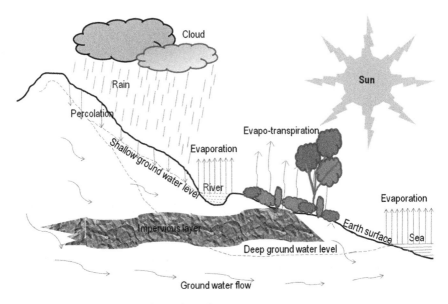

Fig. 1.2 Schematic diagram of the rain cycle

because they are heated by the Sun's rays and form clouds, thus completing a rain cycle as illustrated in Fig. 1.2.

Groundwater may be found close to the Earth's surface or at profound depths. Groundwater is supposed to be wholesome water and rich in minerals as it flows deeper into the Earth's crust. Groundwater is recharged naturally by the percolation of rainwater or surface water.

Regarding groundwater as source of water, the major concerns are depletion and the fast rate of decrease of the water table. In the process of urbanization, the impacts of building development on groundwater hydrology include the following:

1. Increased water demand
2. Increasing dependence on the use of groundwater
3. Over-exploitation of groundwater
4. Increased runoff, decreased groundwater levels, and decreased well yields
5. Decreased open soil–surface area
6. Decreased infiltration and deterioration of water quality

Due to sprawling urbanization, particularly in land-scarce and densely populated areas, the major portion of the urban area remains covered by metal surfaces, such as building, roads, etc., thus leaving few green spaces. As a result, the scope of rainwater infiltration is reduced considerably or even may not occur at all, compared with the rate of withdrawal of groundwater, to meet the growing demand of water mostly in buildings. The natural system established a condition of equilibrium during the process of precipitation, infiltration, surface runoff, evaporation, and evapotranspiration as demonstrated in Fig. 1.2. Continued over-withdrawal of groundwater by pumping leads to a continual decrease of the water table. The water table can decrease at rates ≥ 1 m/year. In Dhaka city, the capital of Bangladesh, at present the average rate of decrease of the groundwater table is approximately 3 m/year [2].

Groundwater is rarely found to be polluted or contaminated, but there are instances of such. Pollution of shallow groundwater is mostly caused by biological activity, much of it due to human intervention. Industries may also play role in polluting groundwater. The other causes may be waste-dumping on the ground or under the ground, land-filling by contaminated soil, and leaching of agricultural pollutants, etc. There is a strong interrelationship between surface streams and groundwater, and if stream water is polluted or tainted, the effects can be transmitted to the groundwater. Pollution or contamination of deep groundwater might also happen due to unethical and unregulated human interventions by deep wastewater-dumping. In coastal areas, the groundwater is often found to be brackish or saline due to the intrusion of saline water from the sea.

In areas where both the surface waters and the groundwater reaches such conditions, at which point they cease to be dependable sources of water, rainwater is considered to be the next best option as a source of water in areas where it is found to be reliable [3].

1.4 Rainwater Utilization

When dependency on water from the usually available sources becomes considerably scarce, then, rainwater is considered to be a potential supplementary alternative option for the water supply in buildings. Rainwater has proven to be the alternative source of water, available at the building door step, posing the lowest risk of use. It can provide a readily accessible and reasonably reliable source of water to meet the demand of water that has fallen short in a building.

Nonpotable uses of rainwater have a wider range and divergence. These may include flushing of water closets and urinals inside buildings; watering indoor plants and irrigating landscapes; outdoor washing such as washing cars, building facades, sidewalks, and roads, etc.; fire suppression such as fire trucks, fire hydrants, sprinkler systems etc.; and supply for chilled water cooling towers, replenishing and operation of water features, water fountains, and laundry if approved by the local authorities. Replenishing of swimming pools may also be acceptable if special measures are taken as approved by the concerned regulatory authority. Rainwater can even be used confidently for drinking and culinary purposes after proper treatment and disinfection.

As a source of drinking water, rainwater is reliable, irrespective of availability and quality, in many parts of the world. The reliability of rainwater resulting from a catchment is based on the probability that the capacity of the system, i.e., the volume of rainwater collected, exceeds or equals the demand or consumption. This system can be set up independently for every household, building, and building complex. The maintenance of the system is simple and can be easily performed by the users.

At present, the interest in using rainwater has increased manifold globally to remedy the scarcity of freshwater primarily due to global warming, heavy pollution of surface waters, nonavailability of water, and depletion of groundwater sources. In many parts on the globe and also in Bangladesh, rainwater is being utilized. In hilly areas of Bangladesh, such as the Kaptai, Bandarban, and Hilltract areas as well as in southern districts, people use rainwater in indigenous ways to mitigate the demand for water for household, irrigation, and navigation purposes. The tribal people of the northern and eastern hills also practice the same.

The feasibility of any system of works depends on its wide acceptance, economy, technological simplicity, and, most importantly, the availability of rain. In this regard, rainwater-utilizing techniques can be considered very feasible. The cost of a rainwater-harvesting system is basically governed by the size of the tank used for storing the rain water. The size of the tank again depends on the diversity and the period of using rainwater in a building. Regarding rain or rainfall, the general concept for determining the feasibility of a rainwater-utilization system is that the rainfall should be >50 mm/month for at least half a year or 300 mm/year unless other sources, such as surface waters and groundwaters, are extremely scarce [4].

1.5 History of Rainwater Harvesting

Rainwater utilization is not a new concept but rather a very old practice. Rainwater utilization started as far back as the emergence of the human race. According to Peterson (1999), as cited in Rao and Giridhar MVSS (2014), the oldest examples of rainwater harvesting found, are associated with the early civilizations of the Middle East and Asia. It was practiced as early as 4500 BC by the people of Ur, a Sumerian city-state in ancient Mesopotamia of Iraq during the Ubaid period (6500–3800 BC) and also latest by the Nabateans and other people of the Middle East [5].

According to Evanari et al. (1971), as cited in Shadeed and Lange (2010), extensive rainwater-harvesting apparatus existed 4000 years ago in the Negev Desert of Israel. According to Evenari et al. (1961), as cited in Grabhoff and Meyer (2013), in the Negev desert runoff from hillsides was collected and stored in cisterns for agricultural and domestic purposes. In ancient Rome, residences were built with individual cisterns and paved courtyards to capture rainwater [6].

According to Hassell, as early as 1700 BC, a sophisticated rainwater-collection and -storage system was built in the Palace of Knossos in the Mediterranean region [7]. This might be the first example of rainwater harvesting in a building. Approximately 3000 years ago (850 BC), King Mesha of Moab in Jordan commanded that cisterns be dug out by every family in the city Qerkhah for themselves [8]. At least 2000 years back, in northern Egypt, tanks ranging from 200 to 2000 m^3 were used for rainwater harvesting, many of which are still operational [9].

The world's largest cistern is probably the Yerebatan Sarayi, in Istanbul, Turkey, built by Caesar Justinian in 527–565 AD. It is 140 m long, 70 m wide, and could store ≤80,000 m^3 of rainwater [10]. This huge structure is completely underground and constructed in a series of intersecting vaults.

From the 6th century BC onward, many settlements in Sardinia, the second largest island in the Mediterranean sea, collected and used roof runoff as their main source of water [11]. Many Roman villas and cities are known to use rainwater as the primary source of drinking water and for domestic purposes.

Examples of rainwater harvesting can be traced dating back to approximately the 9th or 10th century in Asia. According to Prempridi and Chatuthasry (1982), dating approximately 2000 years ago, in Thailand, documents were been found describing rainwater collection from the eaves of roofs or by way of simple gutters flowing into traditional jars and pots [12]. Rainwater harvesting has long been practiced in the Loess Plateau regions of China. Recently, i.e., between 1970 and 1974, approximately 40,000 storage tanks were constructed to store rainwater and stormwater runoff [13].

In India, during the Harappan period (5000–3000 BC), there was very good system of rainwater harvesting and its management [14]. Until 3000 BC, rainwater harvesting happened naturally because rain collected in rivers and natural depressions, from which water was collected and used. At that time, dams of stone rubble were constructed for impounding water [15]. From 3000 to 1500 BC the Indus-Sarasvati civilization grew up when several reservoirs had been built to collect rainwater runoff [15]. In Tamil Nadu, ancient communities stored rainwater separately for drinking versus bathing and other domestic purposes [16].

According to Roy M. Thomas, there is evidences showing the utilization of harvested rainwater in many areas around the world including North Africa (Shata 1982), East and Southeast Asia (Prempridi and Chatuthasry 1982), Japan, China (Gould and Nissen-Peterson 1999), Pakistan and much of the Islamic world (Pacey and Cullis 1986), sub-Saharan Africa (Parker 1973), Western Europe (La Hire 1742; Hare 1900; Doody 1980; Leggett et al. 2001a), North and South America (McCallan 1948; Bailey 1959; Moysey and Mueller 1962; Gordillo et al. 1982; Gnadlinger 1995), Australia (Kenyon 1929), and the South Pacific (Marjoram 1987) [17].

1.6   Rainwater Harvesting Around the World

In recent years, rainwater harvesting, as an alternative source of water, has become popular in many countries. In those countries, rainwater harvesting is being practiced as a mitigation option for persistent water crisis due to insufficient rainfall. The experiences of rainwater-harvesting practices in various countries are discussed below.

1.6.1   Rainwater Harvesting in Bangladesh

The history of rainwater harvesting in Bangladesh is still unknown. However, it can be presumed that it had been practiced in many parts of Bangladesh, particularly in the southern parts, when starting establishing settlement in areas where there was persistent dearth of sweet water due to this country's spatial location near the sea. In those areas, it is still being practiced by collecting rainwater primarily for drinking and culinary purposes. Later, after discovering the presence of arsenic in shallow groundwater in many areas of Bangladesh, rainwater was suggested as the best option for avoiding shallow tube-well water, which had been found to be contaminated with arsenic. In this way, rainwater harvesting became popular and was mostly practiced in rural areas. Harvested rainwater is mostly used for drinking and cooking purposes. The harvested rainwater was found to be free from bacterial contamination for 4–5 months. Since 1997, approximately 1000 rainwater-harvesting systems have been installed in Bangladesh, mostly in rural areas [18]. In the urban setting, the need for rainwater harvesting was first felt in Dhaka city. In 1998, the Dhaka Water Supply Authority, which is responsible for mitigating the water needs of Dhaka city, set an example by installing a rainwater-harvesting system in its head office building at Kawran Bazar. Two ground recharge wells, 750 mm in diameter and 12 m deep, were installed, and the stored rainwater was used for general purposes other than drinking. The system worked for 2 years. Later the system was left abandoned due to construction to build a vertical extension of the building. After this, rainwater-harvesting systems were installed in very few institutional buildings.

1.6.2   Rainwater Harvesting in India

India possesses many draught-prone and water-scarce areas. As a result, in urban centers of those areas, rainwater-harvesting systems have been introduced through supplementing the piped water supply system. To avoid groundwater depletion in the state of Tamil Nadu, the rainwater-harvesting movement launched in 2001. Amendments were made to the Tamil Nadu District Municipalities Act 1920 and Building Rules 1973 making it mandatory to provide rainwater-harvesting structures in all new buildings [19]. In Pune of Maharashtra, rainwater harvesting is made compulsory for any new society to be registered [20]. In Kerala, there occurs periods of severe water scarcity between February and mid May every year, and during summer there are shortages of drinking water. Therefore, the government of Kerala included rainwater-harvesting structures in all new constructions [21]. In Mumbai, The state government has made rainwater-harvesting mandatory for buildings to be constructed on plots sized >1000 m^2 [21]. In Kanpur of Uttar Pradesh, rainwater harvesting has been made mandatory in all new buildings covering an area ≥500 m^2 [20]. In Hyderabad of Andhra Pradesh, rainwater harvesting has been made mandatory in new buildings with an area ≥ 300 m^2 [21]. The state of Rajasthan, located in the northwest of India, experiences a very low annual rainfall of <400 mm and is mostly dry desert. In this region, rainwater harvesting was a traditional practice that has now been revived by extensively practicing it for the past 20 years. The state government has made rainwater harvesting mandatory for all public establishments and all buildings on plots >500 m^2 in urban areas [21]. Failing to do this, water supply can be disconnected. In Gujarat, in buildings of area between 500 and 1500 m^2, installation of rainwater harvesting must be performed per the guideline of the concern development authority. For buildings having area between 1500 and 4000 m^2, one underground recharge well shall have to be installed for every 4000 m^2 or part thereof of building unit [21].

The other places where the installation of rainwater harvesting system has been made mandatory include New Delhi, Bangalore, Surat, Nagpur, Indore of Madhya Pradesh, Haryana, Chandigarh, Himachal Pradesh, Daman and Diu, Goa, Lakshadweep, Meghalaya, Nagaland, Pondicherry, West Bengal, Arunachal Pradesh, Andaman and Nicobar, Orissa, Rajkot, Gwalior, Jabalpur, Ranchi, and Mussoorie [21].

1.6.3 Rainwater Harvesting in China

Although China has enormous water resources, they are unevenly distributed both spatially and temporally throughout the country [22]. In China, 17 provinces have been practicing the rainwater-utilization technique for managing the drinking-water crisis. Gansu is one of the driest provinces in northwest China where the annual rainfall is approximately 300 mm [23]. In this province, rainwater is harvested for supplying as drinking water and for irrigation. In 1995–1996, "1-2-1" rainwater-catchment projects were implemented by the Gansu Provincial Government. The project included one rainwater collection field comprising a roof and a hardened courtyard having area of approximately 100 m^2, two underground water tanks each with capacity of 15–20 m^3, and one piece of land close or near to the household that needs to be irrigated by the harvested rainwater [24]. In Gansu Province, 2,183,000 rainwater-storage tanks were built before 2000 having a total storage capacity of 73.1 million m^3 [25].

1.6.4 Rainwater Harvesting in Africa

In some parts of Africa, the rapid expansion of rainwater-harvesting systems has increased in recent years. The municipal water supply in African cities is accomplished from surface water in dammed rivers or from accumulated rainwater [26]. According to Ngigi, currently in Kenya rainwater-catchment systems are being considered a viable and important water-supply option [27]. Rainwater is also harvested to mitigate the water crises in various other parts of Africa such as Botswana, Togo, Mali, Malawi, South Africa, Namibia, Zimbabwe, Mozambique, Sierra Leone, and Tanzania.

1.6.5 Rainwater Harvesting in Japan

In Tokyo, rainwater harvesting is being promoted to mitigate the shortage of water, control the severity of, and ensure the availability of water during emergencies. Many new public buildings have introduced rainwater-harvesting systems in this city. By this time in Tokyo, >750 private and public buildings have introduced rainwater-collection and -utilization systems [28]. Throughout Japan, rainwater harvesting in buildings and rainwater utilization for various purposes is expanding in both the public and private sectors.

1.6.6 Rainwater Harvesting in Germany

In Germany, rainwater harvesting is gaining popularity. Interest is growing in its promotion, particularly at the level of local government. The reason behind this growing interest in rainwater harvesting, particularly in building, is primarily due to air pollution mainly caused by industries as well as maintaining strict regulations regarding drinking water–quality standards. The use of rainwater in Germany has been limited to various nonpotable uses such as toilet-flushing, pavement-washing, garden-watering, etc. In many cities in Germany, an annual rainwater-drainage fee is charged, which is waived if rainwater is collected or recharged into the ground, thus achieving significant conservation of water. The concerned authorities of those cities provide sufficient grants and subsidies to encourage the development of rainwater-harvesting systems including groundwater recharging. In Berlin, rainwater-harvesting systems have been incorporated into the process of a large-scale urban redevelopment with a view toward controlling urban flooding, saving city water, and creating a better microclimate of the city. In Berlin, approximately 2430 m^3/year of potable water is saved in underground reservoirs through rainwater harvesting [28].

1.6.7 Rainwater Harvesting in the United Kingdom

In the United Kingdom (UK), rainwater harvesting was a traditional way of collecting water for domestic uses. The UK is now under severe water stress, particularly in the south and east of the country. It is believed that situation will get worse with the predicted drier summers and increasing development planned for these areas. As a result, in recent years, rainwater harvesting has undergone a revival in the UK and is becoming more common due to increasing water prices. In the UK, rainwater harvesting is encouraged to employ harvested rainwater for nonpotable uses such as flushing toilets, washing clothes, watering garden, washing cars, etc. Approximately 55 % of treated main water is used for household purposes in the UK, and it is believed that 25 % of household water consumption can be reduced by using rainwater [29]. Because one of the strong regulatory measures to reduce water use, the UK Government has been promoting rainwater harvesting through providing pecuniary incentives, most importantly in the commercial sector. Businesses enterprises that install approved rainwater-harvesting systems are allowed to write off the cost against their taxes [29].

1.6.8 Rainwater Harvesting in Australia

In Australia, rainwater harvesting is gaining attention in the water-management sector due to greater environmental awareness among the population after experiencing long-lasting drought in the past. At present in Australia, various subsidies have been offered by the various government authorities concerned to convince people to resort to rainwater harvesting and to implement rainwater harvesting on a large scale. In urban areas, where a piped water–supply system exists, harvested rainwater supplements for various nonpotable uses of water such as flushing toilets, watering gardens, and washing cars, floors, and pavement, etc. However, in rural and periurban areas where a piped water supply has yet to be developed, harvested rainwater is used as a source of water for almost all variety of purposes including drinking, washing, gardening, etc. [30]. In Southeast Queensland, all new residential buildings must install a rainwater-harvesting system [31].

1.6.9 Rainwater Harvesting in the United States of America

Droughts and water supply–related emerging issues have prompted to consider legislation legalizing the practice of rainwater harvesting for use in households and for lawns in many states. States those have passed legislation to regulate the practice of rainwater harvesting include Arizona, Colorado, Illinois, North Carolina, Ohio, Oklahoma, Oregon, Rhode Island, Texas, Utah, Virginia, Washington, U.S. Virgin Islands, and Hawaii.

Arizona offered a tax credit for water conservation including rainwater harvesting. Colorado allows the collection of rainwater for nonpotable uses. In Illinois, green infrastructures are being developed under the auspices of the Clean Water Act, which emphasizes the conservation of water wherein the incorporation of rainwater harvesting is one of the major concerns in this regard. North Carolina directed for the development of best-management practices for community water efficiency and conservation that include harvesting rainwater. Ohio allows rainwater harvesting, even for potable purposes. A 1979 study showed that approximately 67,000 cisterns existed in the state of Ohio alone [32]. Oklahoma initiated pilot programs that include information campaigns on capturing and using harvested rainwater. In pursuance of the Senate bill, the Oregon Building Codes Division included various guidelines on rainwater harvesting for new and existing repaired buildings for both potable and nonpotable uses of harvested rainwater. In Rhode Island, a tax-credit provision has been created for the installation of cisterns on property to collect rainwater for use in a home or business. In Texas, rainwater has been harvested for nonpotable uses of water. Later, the potable use of harvested rainwater was allowed. In Texas, there is provision for connecting rainwater-harvesting systems, incorporating both filtration and disinfection systems, to public the water-supply system for supplementing potable uses of water. All municipalities and counties in Texas encourage the promotion of rainwater-harvesting systems in all residential, commercial, and industrial buildings through offering various incentives such as giving discounts on purchasing rain barrels or providing rebates for water-storage facilities. In Utah, the owner or lease holder of any land can harvest rain or stormwater directly from the land and store it for subsequent use. In Senate Bill 1416 of Virginia, provision of an income tax credit has been maintained for individuals and corporations those will be installing rainwater-harvesting systems in their buildings or projects. In Washington, state law allows counties to decrease the rates for stormwater–control facilities that utilize rainwater through harvesting it. Since 1964, the U.S. Virgin Islands has had a rule regarding rainwater harvesting that makes the building owners or developers obligatory to incorporate self-sustaining potable-water systems such as rainwater-harvesting systems [33]. In the Island of Hawii, rainwater-harvesting systems have been built to supply water [32].

1.7 Building

A building is an infrastructure built permanently or semipermanently for the use of various types of occupants in safe, comfortable, and good physical and environmental condition. Depending on the nature and size of occupancies, the volume, features, and character of buildings become different. As such, buildings may be categorized in the following ways depending on the nature of their occupancies.

1. Residential
2. Educational
3. Institutional
4. Health care
5. Assembly
6. Business and mercantile
7. Industrial

8. Storage
9. Hazardous, etc.

To live comfortably in good physical and environmental conditions, water is the most important requirement, and it should be of acceptable quality and available in sufficient quantity.

The amount and rate of consumption of water in buildings of various occupancies also varies. The purpose of using water in those buildings also varies considerably. Therefore, depending on the type of use, amount and the rate of consumption, the volume and quality of water can also be varied for saving energy toward achieving sustainability of water management in buildings.

In urban areas, buildings are the major infrastructure for which water is drawn from the main, which is a part of the water-distributing pipe network developed by the water-supply authorities. In an urban water-supply system, river or lake water and underground water are generally used as sources of water. Extraction of sufficient usable water from these sources is becoming either costlier or limited; thus, rainwater is being considered as a dependable alternative source of water for buildings.

Buildings, which are a vital element of the landscape, are built in various forms and shapes from single to multiple stories. The approach to developing rainwater-harvesting systems in very large and high-rise buildings would be different from those of low-rise and small-sized buildings.

1.8 Rainwater Use in Buildings

It is well said that water is life. Water is life not only for living beings but also for nonliving structures such as buildings. Without water, a building is just like monument. Water is required not only during the use of building but also during construction, maintenance, and even in demolition. Water is required not only for consumption by its users, but it is also needed to keep the building clean and its users safe and comfortable. Rainwater harvesting is proven to be an ancient technology that has been adopted to supply water to human settlements and more recently to buildings [34].

The amount of water needed in a building is therefore a major concern in building development. The amount will vary depending on the size, occupancy type, population, and diversification of water use in the building.

The collection of required water is the next concern. The nearest sources of water should be determined. Generally a piped water supply by the water-supply authority is availed as the primary source of water for buildings. When this piped-water source is found to be insufficient for the building, an independent water source is considered by installing a deep tube-well pending permission from the water-supply authority. When both sources are found to be limited or not available, an alternative source is considered, which is rainwater available on and around the building. The rainwater-collection system must be well planned, designed, and maintained.

The greatest challenge is the storage of water, including rainwater, in the building. How large should the storage-facility be, and where should it be located? For the best answer, water-related issues for the building should be well determined before conceiving the building blueprint and be well planned accordingly during the design of the building.

There shall be a rainwater supply system in the building. A rainwater-supply system may be exclusive and independent, or it can be developed in conjunction with the normal water-supply system. The hydrology for a rainwater supply is as same as the basic hydraulic principles of the normal water supply.

It is nearly impossible to meet a building's total demand for water exclusively with rainwater. However, the need of water for a particular demand may be fulfilled by exclusive use of the rainwater. For such particular type of demand, rainwater should be conditioned or treated, stored, and supplied accordingly.

The success and sustainability of using and harvesting rainwater in and around a building depends entirely on careful planning, designing for optimum sizing, choosing appropriate material, and proper installation of the elements used in rainwater-harvesting system. Last, all of these elements must be well maintained. All of these issues related to rainwater harvesting in a building and its premises are discussed in detail in subsequent chapters.

1.9 Establishing a Rainwater-Harvesting System

Establishing a rainwater-harvesting system in a building development requires various efforts in terms of motivating and encouraging the building owners or developers. To become motivated and encouraged to build such a system, the building owner and/or developer must feel obligated or compensated in some way for incorporating the system, which can be achieved through enforcing rules and regulations and enabling practices by performing various mandatory and promotional activities as follows:

1. Formulating policy
2. Creating awareness
3. Instituting incentives
4. Enforcing rules and regulations
5. Formulating codes and guidelines
6. Building capacity to act

1.9.1 Formulating Policy

Majority countries have their own national water policy, water laws, or water bylaws for the sustainable management of their respective water resources. Not all of those policies may address rainwater harvesting at the time of formulation due to not experiencing or failing to anticipate a future water crisis. The need for a policy framework for rainwater-harvesting systems arises mainly when water-policy statements do not focus on this issue [35]. To formulate the necessary laws and bylaws to implement a rainwater-harvesting system, a policy framework regarding rainwater harvesting will be necessary as a guideline. For promoting rainwater harvesting in a building, the policy should focus on building awareness, awarding incentives, and establishing penalties as well control the negative impacts of installing such a system. Implementation of such policy initiatives should be strengthened through legislation. Consideration should also be given to subsidizing facilities or offering incentives for installing rainwater-harvesting systems.

1.9.2 Creating Awareness

Building owners or developers are supposed to possess knowledge of conventional systems of water supply, in which water is generally collected from the supply main or any other independent conventional source of water such as underground water. Against this backdrop, the introduction of a new system of collecting and using rainwater may not be readily accepted due to fastidious notions about the quality of rainwater. To eliminate negative attitudes toward rainwater, the awareness of utilizing rainwater must be enhanced by concerned professionals, government and nongovernmental organizations, and policy-makers alike in order to realize the potential of rainwater harvesting for environmental benefits, human well-being, and ecosystem productivity. Concerned water-service providers must take a leading role in this regard. With the help of all media groups, they can become involved in increasing awareness and implementing rainwater harvesting, predominantly in buildings. The concept of rainwater harvesting can be introduced in academic curricula where applicable.

1.9.3 Instituting Incentives

It is inevitable that incorporation of a rainwater-harvesting system in a building will involve extra initial investment for the purpose which an owner or developer may not be willing to invest unless they are supported by some sort of incentives. Private financial institutions may not agree to provide loans with comparatively less interest rate of return. In these circumstances, financing by concerned government organizations play an important role in raising awareness of its pervasive impact on sustainable economic development and human welfare. Government-owned financial institutions can extend financial assistance in the form of loans or grants to promote rainwater harvesting. In addition, water-service providers may offer some financial benefits in terms of decreased water tariffs. Municipal authorities may offer some sort of tax rebate or exemption for developing rainwater-harvesting systems in buildings. In this way, the payback period of investment for the installation, operation, and maintenance of a rainwater-harvesting system in buildings might be shortened.

1.9.4   Enforcing Rules and Regulations

Even in the midst of various awareness-raising programs and incentives promoting the installation of rainwater-harvesting systems in buildings, people may not be willing to support such a system. In this situation, sufficient rules and regulations should be enacted to compel owners or developers to install rainwater-harvesting systems in buildings where necessary. Necessary rules and regulations regarding the development of a particular rainwater-harvesting system in buildings must be introduced in the concern building development or construction rules. To smoothly promote rainwater harvesting in buildings, the development of laws, bylaws, rules, and regulations must be suitably formulated or modified when necessary. There is a need for a building development–regulatory authority to monitor the abidance of these regulations.

1.9.5   Formulating Codes and Guidelines

Creating awareness, providing incentives, and enforcing rules and regulations will not be sufficient to ensure establishment of rainwater-harvesting systems in buildings in places where rainwater is the reliable alternative option to mitigate a shortage of water. It will be necessary to enable building developers to develop rainwater-harvesting systems in the buildings. For this purpose, exclusive guidelines and codes of practice will be required. The building codes generally furnish the plumbing system by addressing normal water supply, sanitary water supply, and storm drainage only. Building codes or plumbing codes seldom address exclusive and comprehensive guidelines for rainwater harvesting in buildings. In building or plumbing codes, provisions for specific rainwater-harvesting guidelines must be developed to address public-health concerns by stipulating water quality and cross-contamination requirements in particular. Codes on rainwater harvesting should also establish acceptable uses of rainwater, storage requirements, and corresponding treatment options.

1.9.6   Building Capacity to Act

The objectives of establishing rainwater-harvesting system in buildings may not be fulfilled to reach the desired goal in this regard, even though all of the measures mentioned so far have been addressed. This is due to the lack of capacity to plan, design, install, operate, and maintain the system properly. The lack of capacity arises from a lack of knowledge regarding rainwater harvesting, particularly for buildings, which is relatively more oriented toward and associated with the plumbing practices in buildings. In underdeveloped and developing countries, plumbing works are seldom developed properly according to the building codes mostly due to a lack of knowledge, a lack of capacity, and the absence of a regulatory system in this regard. In these situations, additional works for rainwater harvesting in the building may have the same fate, for which the capacity-building tools of the concerned professionals—including architects and engineers, plumbing contractors, and plumbers—is a must. Capacity can be built up through formal education and training program at all technical institutions.

1.10   Challenges to Rainwater Harvesting in Buildings

Although rainwater harvesting is recognized as an age-old traditional technology, its application in today's buildings gains various new dimensions in the present urban context of changed environmental scenarios, continued technological development, and diversified building features. Popularizing rainwater harvesting in buildings is not as easy as was originally believed. Many challenges are to be faced for its widespread implementation as an integral part of the building-development process in the areas of water scarcity. Following are the major challenges to be overcome in this regard.

1. Changing perceptions regarding the use of rainwater.
2. Address conflicting and missing guidelines in codes.
3. Standardize water quality–testing protocols and procedures.
4. Achieve good governance in water-supply authorities.

1.10.1   Changing Perceptions

The general perception regarding rainwater is that it is not safe for drinking. This might be due to the fact that the rainwater, when seen collected or flowing, has a considerable amount of dirt materials in it. Furthermore, it is believed that the surfaces on which rain falls and flows is not free from contaminants and thus it is not possible to obtain rainwater safe for use especially for drinking or cooking. The reason behind such thinking is due to equating rainwater harvesting with storing rainwater in comparatively large containers placed under gutters or collecting rainwater through pipes and directly using it (i.e., without treatment).

Another common perception is that installing a rainwater-harvesting system in a building is costly or is not cost-effective. This notion stems from comparing the cost of developing a harvesting system with the price of supplied water. The availability of water to the main supply and its future scenario is rarely considered. It should be kept in mind that there might be water around, but the cost of supplying water is becoming costlier and the availability more limited every day all over the world.

1.10.2   Conflicting and Missing Guidelines in Codes

For rainwater harvesting in buildings, a separate code of practice is essential. Such a code on rainwater harvesting in buildings will primarily focus on the collection and storage of rainwater in buildings and recharging the groundwater on the building premises only. Other subjects that can be addressed in the code are conditioning, distributing, and draining excess rainwater. There might be one comprehensive code on rainwater harvesting in buildings including all of the aspects related to harvesting. Another code of practice addresses water supply, wastewater, and storm-water drainage, along with other fluid supply and waste disposal, in a building, which is "plumbing" code. In a building, these two codes cannot be practiced side by side and separately. Furthermore, the similar, or more or less similar, jobs of these two systems cannot be developed independently. In some cases, these components must be combined to develop a unified system integrating the plumbing and harvesting components. For example, consider the manner of integrating water-supply piping with the harvested-rainwater supply for flushing a toilet as shown in Fig. 8.2 of Chap. 8. If this job is performed according to two separate codes, one for plumbing and one for rainwater harvesting, there might be chance of missing the guidelines on how to integrate the two systems. Not only might some guidelines be missing, but there might be every chance of also having conflicting guidelines. In this regard, it is preferred to develop the code on rainwater harvesting in a building in coordination and conformity with the relevant plumbing code in practice.

1.10.3   Standardizing Water Quality and Water-Testing Protocols

It is quite obvious that all countries have their own standards of water quality, which are more or less alike with few exceptions and little variation in the acceptable value of some parameters. In standardizing the acceptable value of water parameters, the source is considered to be either surface or groundwater; rainwater is rarely considered. Rainwater might have some elements absent e.g., calcium or magnesium, and contain some additional chemicals, which may not be injurious to health. As a result, there might be misunderstanding in accepting rainwater as potable water. Therefore, in standardizing values of water-quality parameters, all possible elements that might be present for using rainwater shall have to be addressed, and the water quality–testing protocol must be designed accordingly [35].

1.10.4 Absence of Good Governance

The concerned water-supply authority must be committed to fulfilling the demands of their customers by providing wholesome water at a reasonable price by developing a sustainable water-supply system. In the supply system, surface water is considered to be the major source of water, whereas underground water is considered as a supporting source in unavoidable cases. For surface water, an additional cost is involved for treating and conveyance, which increases the price of water depending on the pollution potential in the surface water and the distance of conveyance of water. To keep the price of water to a minimum, the authorities may compromise in terms of the quality or quantity of water and the standard of the supply system, which is undesirable. In some cases, to find underground water that is more pure and more easily available, authorities opt for extracting underground water and keep on installing deep tube wells for subsequent extension of the water-supply network ignoring the adverse impact of over-extraction. Environmental regulatory authorities are considered inactive in this regard, and they are also blamed for failing to maintain surface-water quality. Water-supply authorities, particularly in underdeveloped countries, are mostly found lagging, both in production and supply, and behind the ever-increasing demand of a growing number of customers. Against this backdrop, opportunities for unethical practices are created by customers in connivance with the unscrupulous staff of the supply authorities. Managing illegal water connections, bypassing the water meter, etc., are very common illegal practices. Where water is made easily available through adopting undue measures, establishing rainwater harvesting will be more difficult there.

References

1. Pollution News (2009) Water pollution in Buriganga. Retrieved on 31 May 2015 http://bdnature1.blogspot.com/2009/05/water-pollution-in-buriganga.html
2. Rahman MM, Quayyum S, Sustainable water supply in Dhaka city: present and future. Bangladesh University of Engineering and Technology BUET, Dhaka
3. Haq SA (2005) Rainwater harvesting: next option as source of water. In: Proceeding of 3rd annual paper meet and international conference on Civil Engineering. IEB, Dhaka

4. Worm J, van Hattum T (2006) Rainwater harvesting for domestic use. Agromisa Foundation and CTA, Wageningen, The Netherlands, p 18

5. Sivanappan RK (2006) Rain water harvesting, conservation and management strategies for Urban and rural sectors. National Seminar on Rainwater Harvesting and Water Management, Nagpur

6. Gerston J (1997) Rainwater harvesting: a new water source. In: Texas water savers, news of water conservation and reuse, Vol 3, Number 2. Texas, USA. Retrieved on 31 May 2015 http://twri.tamu.edu/newsletters/texaswatersavers/tws-v3n2.pdf

7. Hasse R (1989) Rainwater reservoirs above ground structure for roof catchments. Gate, Vieweg, Braunschweig/Weisbaden, Germany. Retrieved on 05 Jan 2016 https://englishfc25.wordpress.com/2013/10/29/43/ (cited in Englishfc25)

8. Rain-barrel (2005) The history of rainwater collection. Retrieved on 31 May 2015 http://www.rain-barrel.net/rainwater-collection.html

9. United Nations Human Settlements Programme (UN-HABITAT), Rainwater harvesting and utilisation. Blue Drop Series, Book 2: Beneficiaries & Capacity Builders, pp 10–11, www.unhabitat.org

10. United Nations Human Settlements Programme (UN-HABITAT), Rainwater Harvesting and Utilisation. Blue Drop Series, Book 3: Project Managers & Implementing Agencies, p 10. Visited on 05 Jan 2016 http://www.hpscste.gov.in/rwh/2060_alt.pdf

11. ENGLISHFC25 (2013) History of rain water harvesting. Visited on 01 Jun 2015 https://englishfc25.wordpress.com/2013/10/29/43/

12. UNEP/IETC (1998) Alternative technologies for freshwater augmentation in some Asian Countries (pp 192). The Newzealand Digital Library. Retrieved on 13 Oct 2015 http://www.nzdl.org/gsdlmod?e=d-00000-00—off-0envl—00-0—0-10-0—0—0direct-10—4——0-1l–11-en-50—20-about—00-0-1-00-0-0-11-1-0utfZz-8-00&cl=CL2.7.1&d=HASH0189618d3b8bbcb233c8d080.4.3.1&gt=1

13. United Nations Environment Programme (UNEP) (1982) Rain and storm water harvesting in rural areas. Tycooly International Publishing Ltd., Dublin. Visited on 01 Jun 2015

14. Chatterjee SN (2008) Water resources, conservation and management, chapter 18 rainwater harvesting. Atlantic Publishers and Distributors (P) Ltd., Delhi, p 129

15. C.P.R. Environmental Education Centre (CPREEC) Traditional water harvesting systems of India. Ministry of Environment and Forests (MoEF), Government of India. Visited on 01 Jun 2015 http://www.cpreec.org/pubbook-traditional.htm

16. Ishaku HT, Rafee MM, Ajayi APA, Haruna A (2011) Water supply Dilemma in Nigerian rural communities: looking towards the sky for an answer. J Water Res Prot, 3: 598–606. Visited on 01 Jun 2015 http://www.SciRP.org/journal/jwarp, file:///C:/Users/User/Downloads/JWARP20110800001_76891241.pdf

17. Roy MT (2009) Studies on the quality of rainwater at various land use locations and variations by interaction with domestic rainwater harvesting systems. In: A thesis for the award of the degree of doctor of philosophy, Division of Civil Engineering, School of engineering Cochin University of Science and Technology, Cochin-682 022. Visited on 01 Jun 2015 http://dyuthi.cusat.ac.in/xmlui/bitstream/handle/purl/2123/Dyuthi-T0474.pdf?sequence=16

18. Global Development Research Center (GDRC) (2015) Rainwater harvesting and utilisation, an environmentally sound approach for sustainable Urban water management: an introductory guide for decision-makers. Visited on 04 Jun 2015 http://www.gdrc.org/uem/water/rainwater/rainwaterguide.pdf

19. Directorate of Town Panchayet (2014) Rainwater Harvesting. Visited on 04 Jun 2015 http://www.tn.gov.in/dtp/rainwater.htm

20. Quora (2014) What state has made rain water harvesting compulsory for all houses? In: Hoffman S (ed). Visited on 04 Jun 2015 http://www.quora.com/What-state-has-made-rain-water-harvesting-compulsory-for-all-houses

21. Quora (2014) What state has made rain water harvesting compulsory for all houses? In: Singhal S (ed). Visited on 04 Jun 2015 http://www.quora.com/What-state-has-made-rain-water-harvesting-compulsory-for-all-houses

22. Qiang Z, Yuanhong L (Undated) Rainwater harvesting in the Loess Plateau of Gansu, China and Its Significance. Gansu Research Institute for Water Conservancy, Lanzhou, China, Email: qzhu@zgb.com.cn, (P)
23. Centre for Science and Environment. Rainwater Harvesting in Gansu Province, China. Visited on 01 Jun 2015 http://www.rainwaterharvesting.org/international/china.htm
24. Zhu Q (Undated) Rainwater harvesting is an integrated development approach for the untainous areas with water scarcity'XI IRCSA Conference proceedings. http://www.eng.warwick.ac.uk/ircsa/pdf/11th/Zhu.pdf. Retrieved on 7 Jan 2016
25. United Nation Environment programme (UNEP) Rainwater harvesting and utilisation, an environmentally sound approach for sustainable Urban water management: an introductory guide for decision-makers. Newsletter and Technical Publications. Retrieved on 7 Jan 2016 http://www.unep.or.jp/ietc/publications/urban/urbanenv-2/9.asp
26. Racked E, Rathgeber E, Brooks DB (1996) Water management in Africa and the Middle East challenges and opportunities. International Development Research Centre, PO Box 8500, Ottawa, ON, Canada K1G 3H9. Visited on 10 Jan 2016 http://www.idrc.ca/EN/Resources/Publications/openebooks/289-9/index.html
27. Government of Kenya (GoK) (1994) National development plan, 1994–96. Government Printers, Nairobi, Kenya
28. United Nations Environment Programme (UNEP UD) Rainwater harvesting and utilisation: an environmentally sound approach for sustainable Urban water management: an introductory guide for decision-makers. Visited on 10 Jan 2016 http://www.unep.or.jp/ietc/publications/urban/urbanenv-2/index.asp
29. Hassell C, Rainwater harvesting in the UK—a solution to increasing water shortages? 5, Roseberry Gardens, London, N4 1JQ, UK, cath.hassell@ech2o.co.uk
30. Hajani E, Rahman A (2014) Reliability and cost analysis of a rainwater harvesting system in Peri-Urban regions of greater Sydney, Australia. Water, 6: 945–960. doi:10.3390/w6040945 file:///C:/Users/User/Downloads/water-06-00945-v3 %20(1).pdf Retrieved on 10 Jan 2016
31. Localizer E (2013) Australian building codes mandate rainwater harvesting. Retrieved on 15 Nov 2015 http://ecolocalizer.com/2013/09/07/rainwater-harvesting-is-required-in-australia/
32. Stark T, Pushard D (2008) The state of rainwater harvesting in the U.S. Retrieved on 13 Jan 2016 http://www.nesc.wvu.edu/pdf/dw/publications/ontap/magazine/OT_FA08.pdf
33. National Conference of State Legislatures (NCSL) (2012) State rainwater harvesting and graywater laws and programs. Visited on 01 Jun 2015 http://www.ncsl.org/research/environment-and-natural-resources/rainwater-harvesting.aspx
34. Gould J, Nissen-Peterson E (1999) Rainwater catchment systems for domestic supply: design, construction and implementation. Intermediate Technology Publications, London. Retrieved on 13 Jan 2016 http://basharesearch.com/IJSGWM/6010209.pdf (cited in conserving rain water: a revolutionary solution for water scarcity problem! A case study of Lucknow City of Uttar Pradesh, India by Purnima Sharma. International Journal of surface and Groundwater Management, Vol 01, NO 02 Jul 2014)
35. Novak CA, Van Giesen GE, Debusk KM (2014) Designing rainwater harvesting system, Integrating rainwater into building system. John Wiley and Sons, New Jersey, USA

Chapter 2
Rain and Rainwater

Abstract To deal with rainwater harvesting, knowledge of rain, rainfall, and rainwater is necessary. Knowledge of the formation of rain in clouds and its pattern of falling to the Earth's surface helps one to understand more about quality of rain and the quantity of rain that can be collected for harvesting. To collect rainwater at a particular place, the intensity of rainfall to be measured should be known. A rain gauge is used for this purpose. Installation of a rain gauge and the technique of measuring rain is important. In a particular place, the rainfall pattern and its distribution over a period of time is variable. Therefore, rainfall intensity should be studied and recorded for the planning and design of a rainwater-harvesting system. The nature and quality of the rain, and also of the rainwater, also varies from place to place due to pollution of the air in these areas. As a result, for any place where rainwater harvesting is to be performed, the quality of rainwater must be tested or be known from secondary sources. With a view to enriching the concept of rainwater harvesting in buildings, details about the formation of rain, pattern of rainfall, rainfall intensity, global rainfall scenario, and causes of degrading quality and nature of rainwater are discussed in this chapter.

2.1 Introduction

To understand rainwater harvesting, it is necessary to know about rain, rainfall, and rainwater. Rain is the key element in the water cycle on the Earth. It is considered to be the root source of all freshwater on the Earth's surface water and underground water and is responsible for the accumulation of most of the water on the Earths' surface and underground. Rain makes the environment clean and healthy by sustaining a healthy niche for diverse ecosystems. Rain is the purest form of water when formed in a cloud but can be contaminated by absorbing suspended matters in the atmosphere while falling to the Earth. After falling on surfaces of various natures on the Earth, some rain percolates into the ground, and the rest flows downward toward any nearest water course. Rainwater harvesting is performed with that portion of rainwater that is yet to reach any flowing water course,

© Springer International Publishing Switzerland 2017
S.A. Haq, PEng, *Harvesting Rainwater from Buildings*,
DOI 10.1007/978-3-319-46362-9_2

absorbed by any surface other than ground, or evaporated. In this chapter, the characteristics and properties, etc., of rain, rainfall, and rainwater are discussed.

2.2 Clouds

Clouds are formed by rising water vapours created from heated waters on the Earth's surface. When warm and moist vapour rises high into the atmosphere, it is cooled by an adiabatic process. The continuous cooling of vapour results in condensation and finally forms raindrops. These droplets move around in the cloud and collide with each other. In this process, the droplets increase in size until they are heavy enough to fall to the Earth's surface.

Clouds generally occur at higher levels of the atmosphere wherever there is sufficient moisture to create condensation. The troposphere is the layer of the atmosphere where almost all clouds are formed. Clouds vary in structure and composition, which are classified into three groups: low-, middle-, and high-level clouds. Low-level clouds range from the Earth's surface to the height of approximately 2000 m; middle-level clouds range from approximately 2000 to 6000 m; and high-level clouds exist above 6000 m [1].

Not all clouds produce rain. Clouds that produce rain usually have particular characteristics. They are often large in size. Initially a white-colour cloud is formed, which eventually transforms to gray and finally appears dark because at this stage clouds become so large and full of water that sunlight cannot shine through. The heaviest rain falls from the deepest, darkest clouds, which are formed at a comparatively very high altitude.

Cumulonimbus or thunderstorms clouds get up to an enormous height. In tropical regions, they can reach up to approximately 20 km above the Earth's surface [2]. These clouds can generate torrential rainstorms, with rain falling at high velocity, as much as 90 cm in an afternoon [3]; they can also create thunder and lightning. The formation of tornadoes is also associated with these types of clouds. Various other rain clouds cause considerable variation in rainfall. Layered clouds, such as the comparatively thinner and lighter nimbostratus clouds, usually cause slower and steadier rainfall that may last for hours and even days sometimes [4].

2.3 Rain

Virtually the product of the condensation of water vapour in the atmosphere, which falls under gravity, is termed as "precipitation". The common forms of precipitation are hail, rain, drizzle, snow, etc. Rain is the liquid form of precipitation, which fall from the sky in droplets.

Warm air rises and carries water vapour from the surface waters and moist surfaces high into the atmosphere. As the vapour rises, it expands in size and becomes cooled. At high altitude in the atmosphere, suspended dust and smoke particles in clouds make places for the vapour particles to settle down and finally form into droplets. The drops formed are nearly spherical in shape because of the surface tension of water. Some vapour freezes into tiny ice crystals, which attract cooled water drops. The water drops finally freeze to form ice crystals, which grow into larger crystals called "snowflakes." As the larger snowflakes grow heavy, they start falling. When the snowflakes come across the warmer air on its way down, they melt and convert to raindrops. In tropical climates, the rising lighter dusts or sea salt particles attract cloud droplets to combine together around them. The droplets and particles collide with each other and grow in size until they become heavy enough to fall down as rain.

Sizes of raindrops vary from 0.5 to 10 mm in diameter [5]. Rainwater droplets >0.5 mm in diameter are termed "drizzle." Larger rain drops are of oblate or pancake-like shape, and smaller raindrops are more or less spherical.

2.4   Rainfall

Rainfall is an important atmospheric phenomenon. It is considered to be the critical source of water falling from the atmosphere on the Earth's surface for the sustenance of all life. Rainfall is characterized by its amount, intensity, and distribution over time.

2.4.1   Amount of Rainfall

The amount of rainfall is generally expressed in millimeters (mm) per day (mm/day), which represents the total depth of rainwater in millimeters, on a solid surface, during a 24-hour day. The amount of rainfall is determined by summing all of the rainfall events that occurred during a 24-hour day. A rainfall of 1 mm generates 0.001 m^3 or 1 L of rainwater on 1 m^2 of a surface. In a day when the rainfall is found to be >1 mm, the day is considered a "rainy day".

2.4.2   Rainfall Intensity

The rainfall intensity is the measure of the amount of rain that falls during a specified period of time. The intensity of rain is determined by measuring the height of the water layer accumulated over a surface in a period of time. For rainwater harvesting, the information regarding rainfall intensity in the area concern is useful.

Rainfall intensity is thus calculated by the depth of rainwater accumulated on a surface in millimeters divided by the duration of the rainfall in hours. It is expressed in millimeters per hour (mm/h).

Example
Consider a rain shower that lasted for 4 h and produced 40 mm of rainwater. The intensity of this rain shower is

$$\frac{40 \text{ mm}}{4 \text{ hours}} = 10 \text{ mm/hour}$$

On the basis of a certain range of rainfall intensity, rain is classified as below [6]:

1. Rainfall <2.5 mm/h is termed "light rain".
2. Rainfall between 2.5 and 7.5 mm/h is termed "moderate rain".
3. Rainfall 7.6 mm/h and 50 mm is termed "heavy rain".
4. Rainfall >50 mm rainfall is termed "violent rain".

Annual Rainfall: In a time period of 12 months, total measured rainfall is the annual or seasonal rainfall, which is generally used for expressing the rainfall scenario of a country.

Rainfall of a season or year is counted with a respective long-period average (LPA) and coefficient of variation (CV). The LPA is the average rainfall for a considerably long period, and variation from this is expressed in terms of coefficient of variation. The seasonal rainfall is classified as excess, normal, or deficient as shown below [7].

1. **Excess rainfall**: If rainfall occurrence is more than the total of the LPA and the coefficient of variation (LPA + CV), then it is termed "excess" rainfall for that year.
2. **Normal rainfall**: When the rainfall occurrence is within the addition and subtraction of the LPA and the coefficient of variation (LPA ± CV), it is termed "normal" rainfall for that year.
3. **Deficient rainfall**: The seasonal rainfall is classified as "deficient" rainfall when the actual rainfall occurrence is less than the subtraction of the LPA and the coefficient of variation (LPA–CV).

The amount of rainfall may also be expressed as "heavy" to "scanty" depending on the annual intensity of rainfall for an area as follows [8].

1. Heavy rainfall: Rainfall intensity >2000 mm/y.
2. Moderate rainfall: Rainfall intensity 1000–2000 mm/y.
3. Less rainfall: Rainfall intensity 500–1000 mm/y.
4. Scanty rainfall: Rainfall intensity <500 mm/y.

2.4.3 Rainfall Distribution

Rainfall does not occur uniformly throughout an hour, the day, the month, or the year. It may fall in segregated patterns for several couple of minutes in an hour, a couple of hours in a day, and a couple of days in a month or year. The duration of gaps between the rain periods is important to know, particularly for rainwater harvesting. Evenly distributed rainfall is better than poorly distributed rainfall throughout the year for rainwater storage. For a long, dry period between rain events, a comparatively large storage tank will be required to store rainwater. Therefore, rainfall distribution over the years should be well observed and determined for making sound decisions about rainwater storage. An example of poorly versus evenly distributed rainfall over a year is shown in Figs. 2.1 and 2.2, respectively.

2.4.4 Rainfall Measurement

The total amount of rainfall during a particular period is measured as the depth of rainwater covering an area without any runoff, infiltration, and evaporation. The depth of rainwater is generally measured and expressed in millimeters. Rainfall is determined by measuring its volume by an instrument called a "rain gauge". Accuracy of rain-volume measurement is primarily governed by the wind, the height of the gauge, and the gauge's exposure to the open sky. The accumulation of rainwater in the gauge is a function of the height of the rain gauge.

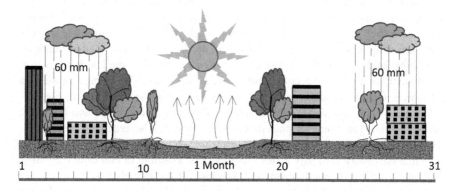

Fig. 2.1 Poorly distributed rainfall (120 mm per month) during a period of 1 month

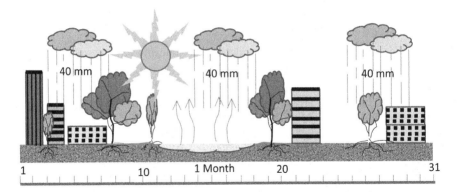

Fig. 2.2 Evenly distributed rainfall (120 mm per month) during a period of 1 month

2.4.5 Rain Gauges

Instruments for measuring the volume or weight of falling rain are called "rain gauges". Rain gauges may be of the recording type and the nonrecording type. Nonrecording types are ordinary rain gauges, which require measurement of rain manually by the observer. Some recording types, such as siphon rain gauges, have a built-in recorder. Other types, such as tipping-bucket rain gauges, have a recorder attached to them, and remote data collection is made possible by setting a recorder at a site distant from the gauge.

2.4.6 Ordinary Rain Gauge

The ordinary rain gauge is basically a receptacle with a tapering funnel, with a fixed orifice diameter, to collect rainwater for measuring. The other important components are a shell, a storage bottle, a storage vessel, and a rain-measuring glass. The rain-measuring glass cylinder is graduated according to precipitation amounts depending on the diameter of the orifice of the receptacle. The shell basically acts as a container for the storage bottle and the storage vessel. The storage vessel is a cylindrical metallic container that houses the storage bottle as shown in Fig. 2.3. A leak-proof collector rim is placed above a funnel, which leads to and fits into a storage bottle. The bottle is placed under the funnel. The receiving area of the collector ranges between 200 and 500 cm^2. The rim of the collector has a sharp edge that falls away vertically inside. To avoid the splashing of rain from the gauge, the wall is made sufficiently deep, and the slope of the funnel is >45°.

Rainwater entering through the receptacle accumulates in the storage bottle, and the precipitation amount is measured with the measuring glass. The measuring

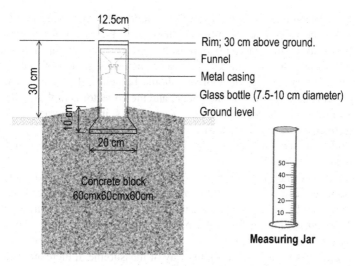

Fig. 2.3 Ordinary rain gauge and measuring jar

cylinder is made of clear glass or plastic. The diameter of the measuring cylinder is not more than one third of the diameter of the rim of the gauge. The graduations are clearly engraved generally in 0.2-mm intervals.

2.4.7 Installing the Rain Gauge

The rim of the rain gauge must be placed truly horizontal. Therefore, the site where the rain gauge is to be placed should be a level ground and the surrounding ground should be uniform. No external obstacles—such as other instruments, building, or trees—should be closer than four times of the height of the gauge. The ground should be covered with grass or be of loose earth. Extremely exposed sites, such as the top of a hill, should be avoided. For extremely exposed sites without any natural shelter, rain-gauge shields are sometimes used. The rain gauge must be firmly mounted on a concrete base. The rain gauge should be installed partly exposed at usually 300 mm above ground level.

2.4.8 Rain Measurement

Rain-gauge measurements are to be taken at the same time generally each morning. A graduated measuring cylinder is used for this purpose. If it rains during observation, the measurement should be taken quickly to avoid unreliability.

To take the reading, rainwater from the storage bottle is poured into the measuring jar and the value read from the graduations. If there is considerable rainfall, the process of taking the reading may must be performed in two or more stages. In multistage reading, the bottom of the water meniscus should be taken as the defining line. When reading, the cylinder should be held truly vertical. The empty storage vessel is then returned to the gauge, and the collector is replaced. Rainwater that overflows from the storage bottle accumulates in the storage vessel. The amount of over flown rainwater is also measured with the measuring glass and the amount added to the amount of precipitation in the storage bottle in order to determine the total precipitation.

If no graduated measuring cylinder is available, a dipstick or rod graduated in cubic centimeters can be used for this purpose. The former is preferred because it makes it easier and quicker to take measurements. The procedure of measuring with a dipstick or rod is more or less the same, but in finding the reading the observed volume is divided by the surface area of the collector of the gauge in cm^2 to determine the amount of rainfall. In this way of measurement, there are the chances of making an error.

2.4.9   Observations

While performing observation, the receptacle is removed first, and the storage bottle is removed to pour the accumulated rainwater into the measuring glass. After taking the measurement, the storage bottle and the receptacle are reset. If the amount of precipitation is found to be too large or if precipitation is in progress, the storage bottle and the vessel are replaced first, and then measurement is performed indoors. If snow or other solid precipitation accumulates in the receptacle, a known amount of warm water to melt it is poured in, and the total amount of water is measured. Then in the measurement, the amount of warm water added is subtracted from the total water. If the amount of precipitation is large, the measurement is repeated and then added to the individual totals obtained.

Rainfall is generally observed in units of 0.1 mm. Readings observed at <0.05 mm is recorded as a "trace" [9]. The "trace" amount of rainfall is kept recorded when there is no or little sign of precipitation appearing in the gauge after a slight rainfall since the last rain-gauge reading.

It is conventional to allocate the 24-hour catch observed in the rain gauge before or at 09.00 h to the previous day [9]. For example, the catch measured at 08.00 h on 1 July will be shown in the record dated 30 June and is included in the June totals. The hour of observation should, however, still coincide with the local practice.

2.4.10 Maintenance

Rain gauges should be checked for leakage and accumulation of dust or leaves. All of the accumulations should be removed periodically from the collector. The inside should be cleaned but not polished. The measuring cylinder should be kept clean and should not be dented. A spare measuring cylinder should be kept available at all times. Nearby plant growth around and above the rain gauge should be checked often and removed if present.

2.5 Global Rainfall Scenario

The surface area of Earth is divided into several very large land masses of seven regions commonly regarded as continents. These continents include Africa, Antarctica, Asia, Australia, Europe, North America and South America. Rainfall varies from place to place on these land masses and also within the land mass. Short and heavy showers are common in warm places, such as the tropics, due to the higher temperatures causing the rapid rise of air, thus creating large rain clouds. The rainfall scenario of the different continents is discussed below.

2.5.1 Africa

A major portion of the northern half of Africa primarily comprises desert or arid land. The Sahara desert is located in this part of Africa. The southern part of Africa possesses savanna plains, and dense forest regions are located in its central portion. The desert regions virtually do not have any rain for several years in a row. Southern Africa receives most of its rainfall in summer. Once a decade, tropical cyclones lead to excessive rainfall across the region. Sub-Saharan West Africa has three climate regions: West Sahel, the Central Sahel, and the Guinea Coast. In East Africa, rainfall occurs during two wet seasons March to May and October to December.

From West Africa to East Africa, the rainfall pattern varies considerably. In the northwest region, annual rainfall often remains <200 mm. In contrast, much of the eastern Highveld receives 500–900 mm of rainfall per year; but occasionally rainfall in this area exceeds 2000 mm [10]. A large area of the center of the South Africa receives an average rainfall of 400 mm, and there are wide variations closer to the coast [10]. On average, <1000 mm of rain falls per year across most of Africa [11].

2.5.2 North America

Like the varied landforms of North America, its climates are also of a varied nature. North America experiences three different climate classifications within its region: arid, humid, and continental. Arid climates are in areas such as east of the Sierra Nevada. Humid climates are in the south of the Great Salt Lakes to portions of eastern Quebec having 750–1500 mm of rain annually [12]. Continental climates are characterized as hotter summers and colder winters, which prevail in much of the interior of North America. The average rainfall in this climate ranges from 250 to 750 mm in a year [12]. The region in this climate classification sits in the transition zone between humid climates and arid climates, which is defined as the 500-mm line of annual precipitation that runs near the 100th meridian. Rainfall on the Great Plains area decreases from east to west [12].

2.5.3 South America

South America is located in the southern hemisphere of the globe. The countries of South America experience four distinct seasons: spring, summer, autumn, and winter. In the north of the continent, the countries experience high temperatures and high rainfall more constantly throughout the year. South America can be broadly divided into seven different climatic zones as follows [13].

1. Desert: Warm to high temperatures with very little rainfall
2. Grassland: Two seasons, summer and winter, with above-average rainfall
3. Deciduous forest: Four distinct seasons with wet winters
4. Rainforest: High temperatures and high rainfall throughout the year. In the Amazon basin rainfall occurs, over 200 days a year on average, and total rainfall reaches as high as 2000 mm/y [14]
5. Savanna: Very high temperatures prevail year round, and rainfall occurs during the summer only
6. Mediterranean: Warm to high temperatures with rainfall in the autumn and winter
7. Alpine mountains: Cold, windy weather and snowy terrain.

In South America, heavy rainfalls are experienced in the Amazon river basin, the coastal parts of French Guiana, Guyana and Suriname, the southwestern parts of Chile and Columbia, and the Ecuador coasts. The rainiest place is Quibdo in Colombia, which receives approximately 8900 mm of rain annually [15].

The desert regions of Chile are the driest part of South America. Coastal parts of Peru are also dry lands. The Atacama Desert in Chile receives a very trace amount of rainfall of approximately 0.1 mm/y [16]. The coastal lands of Peru have humidity level averaging approximately 98 % but receive only 25–50 mm of rainfall annually [17].

2.5.4   Antarctica

Antarctica is the coldest land mass on the Earth. It is also extremely dry, theoretically a desert-like area that receives very poor precipitation throughout the year. Weather fronts rarely reach far into the continent. In Antarctica, almost all of the precipitations falls as snow. The surface waters in these areas mostly remain as ice, and the air is also very cold and holds very little moisture compared with the air of warm regions. As a result, these areas experience very little rainfall. The average precipitation is approximately 166 mm/y [18]. Throughout the Antarctic, the rainfall intensity varies widely from high values, e.g., an average of 1000 mm/y in some coastal regions [18], to very low values, e.g., as low as 50 mm/y in the deep interior [19].

2.5.5   Australia

Among all of the inhabited continents, Australia receives the lowest average annual rainfall. The island continent of Australia features a wide range of climatic zones. The north region can be characterized as the tropical climatic zone; the interiors comprising expanses of dry land can be considered as the arid climatic zone; and the southern part can be characterized as the temperate climatic zone. Australia is considered to be the world's second-driest continent after Antarctica and has highly variable rainfall.

The rainfall pattern is more or less fluctuating and highly seasonal in character with a winter rainfall regime in the south and a summer regime in the north. Comparatively uniform rainfall occurs in much of New South Wales, parts of eastern Victoria, and southern Tasmania. Central Australia is characterized by dry lands where rainfall is rare and unpredictable.

The average (mean) annual rainfall is <600 mm across 80 % of the continent and <300 mm across 50 % of the area [20]. As a whole, in Australia the annual average rainfall is approximately 465 mm [21]. Rainfall generally increases toward the coastal regions of mountainous areas, and higher rainfall occurs in the far north. Much of central Australia receives scanty rainfall. From the long-term average of annual rainfall across Australia, the variation is observed to range from <300 mm in most of central Australia to >3000 mm in parts of far northern Queensland [22].

2.5.6   Europe

Europe comprises lands of relatively temperate areas that receive rainfall almost throughout the entire year. Similarly, areas around the equator are mostly wet all of the year round. In Eastern Europe, there are four seasons, whereas southern Europe experiences two more or less distinct conditions: a wet season and a dry season. Hot and dry conditions prevail during the summer months. The annual precipitation in Europe varies from a maximum of 5000 mm to a minimum of 150 mm [23]. The average precipitation in Europe is considered to be approximately 789 mm/y [23]. In Southwestern Scandinavia and the mountains of Scotland, the Dinara mountains, and Caucasus, precipitation reaches as high as 5000 mm [23]. In continental Europe, the average precipitation is as low as 769 mm [23]. In a study on rainfall in Europe carried for during 1950–2008, the number of rainy days per year did not increase, but the length of wet spells and the periods in which rainfall occurred on a number of consecutive days increased by approximately 15–20 % [24].

2.5.7 Asia

Asia is the largest and most populous continent on the globe. Asian countries are grouped in three zones: Southeast Asia or South Asia, West Asia, and Central Asia. India and Bangladesh are in South Asia zone.

South Asia: South Asia experiences comparatively more rainfall than the rest of the world. Much of the South Asia region receives >1500 mm of rainfall annually, and many areas generally receive double and even triple that intensity of rainfall [25]. The rainfall pattern is distinctly affected by two prevailing air currents: the northeast (or dry) monsoon and the southwest (or wet) monsoon [25]. A review report of >100 research articles showed that with a continuing increase in CO_2 and global warming, South Asia may expect comparatively more rainfall, due to the expected increase in atmospheric moisture, as well as more variability in rainfall pattern [26].

West Asia: West Asia is comprised of the countries in the Middle East and those on the Arabian Peninsula. These lands can be characterised as deserts and plateaus with high atmospheric temperatures and receiving low precipitation. In the mountainous region of Saudi Arabia, annual rainfall has been recorded to be in excess of 400 mm; the highest range is approximately 500–600 mm in 1 year [27]. Almost 90 % of Jordan is desert where annual rainfall is found as low as 200 mm, and the maximum annual rainfall is approximately 600 mm in the northwest corner of the country [28].

Central Asia: Central Asia is characterized by extreme aridity and sparse precipitation. Its remoteness from oceans is the major cause of little atmospheric precipitation. Annual average rainfall ranges between 191 mm in Turkmenistan to 691 mm in Tajikistan [29]. Furthermore, the impact of climate change may result in increased heat stress, drought, and desertification leading to decreased rainfall and runoff.

2.6 Rainfall in India

India is the seventh largest country in the world with a land area of 3,287,263 km^2. The country has diverse landscape experiencing with varying climate. Throughout India, the climate is widely varied with a tropical monsoon climate in the south and a temperate climate in the north. Rainfall distribution in India is therefore uneven and irregular. It also varies in space and time.

The total rainfall scenario shows an increasing trend generally toward the east and at high altitudes. In the Himalaya Mountains, at an elevation of approximately 1500 m, there is a high increase in precipitation. In the northeastern part of the Indian peninsular plateau and the Ganga plain, the monsoon depressions cause widespread rainfall. Rainfall is more or less evenly distributed in the northeastern part of the country. A small village in the Khasi hills of Meghalaya state of India is Mawsynram, which is the wettest place on Earth with an annual rainfall of 11,872 mm [30]. Average rainfall throughout the state varies considerably. The western part, which is desert, receives an annual rainfall of 100 mm. The southeastern part of India receives approximately 650 mm of rainfall annually [31]. Approximately 80 % of annual rainfall in India occurs during the monsoon season, which comprises two monsoons: the southwest monsoon (June to September; spreads to western, northern, and central India) and the northeast monsoon (October to December; occurs mostly in South India). In India, the highest mean monthly rainfall is 286.5 mm, which occurs during July and contributes approximately 24.2 % of the annual rainfall. The average annual rainfall in India is considered to be 1082.2 mm [32]. On the basis of total annual rainfall intensity, four groups of areas have been categorized in India [33]. The areas, with each group being certain range of rainfall, are as follows.

Areas of inadequate or scanty rainfall: These are the areas where rainfall is <500 mm/y. The areas include the Northern part of Kashmir, the Karakoram mountains and the area lying to the north of the Zanskar Himalaya range, Western Rajasthan, Kutch Coast and the delta of the Luni River and touching Haryana near Hisar [33]. A small patch of Punjab, Andhra Pradesh, Maharashtra, Karnataka and Ladakh, Punjab-Haryana plain, Amritsar, western Gujarat, western part of the Maharashtra plateau, Karnataka, and southeast Andhra Pradesh also receive scanty rainfall [33].

Areas of low rainfall: These are the areas where the annual rainfall is within 500–1000 mm/y. The areas, in this group include the Western Ghats mountain range, the interior of Tamil Nadu (excluding the coastal area), the upper Ganga valley, eastern Rajasthan, and Punjab [33].

Areas of medium rainfall: Areas receiving rainfall ranging 1000–2000 mm/y fall in this group. The middle Ganga valley, northeastern peninsular are, Himachal pradesh, Uttar pradesh, Bihar, West Bengal (excluding Darjeeling and Duars), Orissa, and a few districts of eastern Maharashtra are the areas are in this group [33].

Areas of high rainfall: The areas where rainfall is >2000 mm/y are grouped as areas of high rainfall. The west costs on the Western Ghats, Southern slopes of the eastern Himalayas in the northeast, Meghalaya hills. Assam, Darjeeling district of West Bengal, and Arunachal pradesh are the areas of very high rainfall [33].

2.7 Rainfall in Bangladesh

Bangladesh is located near the south of the foothills of the Himalayas. Bangladesh experiences a tropical monsoon climate characterized by wide seasonal variations ranging from moderate to heavy rainfall, high temperatures, and high humidity. Regional climatic differences in this relatively flat country are minor. The monsoons in Bangladesh originate due to large variation between low and high air-pressure areas created by differential heating of land surfaces and water bodies. During the pre-monsoon hot months, the tempered air rises over the Indian subcontinent creating low-pressure areas into which cooler, moisture-bearing winds rush from the Indian Ocean and move to the north. At the end of the monsoon, near the Himalayas, the monsoon winds turn west and northwest.

Considering the climatic condition, three distinct seasons are recognized in Bangladesh: the cool and dry season (November to February), the pre-monsoon hot season (March to May), and the rainy monsoon season (June to October) [34]. In general, the maximum temperature in summer ranges between 38 and 41 °C. April is considered to be the hottest month in most parts of the country, and January is the coolest month when the average temperature range for most of the country is 16–20 °C during the day time and approximately 10 °C at night [35]. Average daily humidity ranges from March having lows between 55 and 81 % to July having highs between 94 and 100 % [35]. This high humidity range characterizes the appearance of the southwest monsoon or monsoon. Approximately 10 % of annual rainfall occurs during the pre-monsoon period; 80 % occurs during the monsoon period; and 10 % occurs during the post-monsoon period [36]. Study have shown that there is positive trend in annual rainfall and a negative trend in winter rainfall in Bangladesh, which indicates that rainfall is concentrating in the pre-monsoon and monsoon periods [37]. The overall trend of the 5-year moving average shows an increasing trend of rainfall in Bangladesh [37]. During the monsoon period, the countrywide mean rainfall is significantly increasing at 74.9 mm/y [38].

In Bangladesh, the higher rainfall occurs in the eastern part of the country, and the highest rainfall occurs in the northeastern region and eastern part of the coastal zone. The coastal and hilly areas have higher rainfall. The average annual rainfall ranges from a maximum of 5690 mm in the northeast of the country to a minimum of 1110 mm in the west [39]. Heavy rainfall causes flooding in almost every year inundating approximately 20 % of the country increasing ≤68 % in extreme years. The spatial distribution of rainfall in Bangladesh is shown in Fig. 2.4. The average coefficient of variation for the annual rainfall is 20.86, and the annual number of rainy days is 13.76 for overall Bangladesh [40]. The value of average precipitation in depth (mm/y) in Bangladesh was found to be 2666 mm as of 2014 [41, 42]. During the past 52 years, this indicator reached a maximum value of 2666 mm in 2014 and a minimum value of 2666 mm in 1962 [42], i.e., approximately 2.67 cum of rainwater

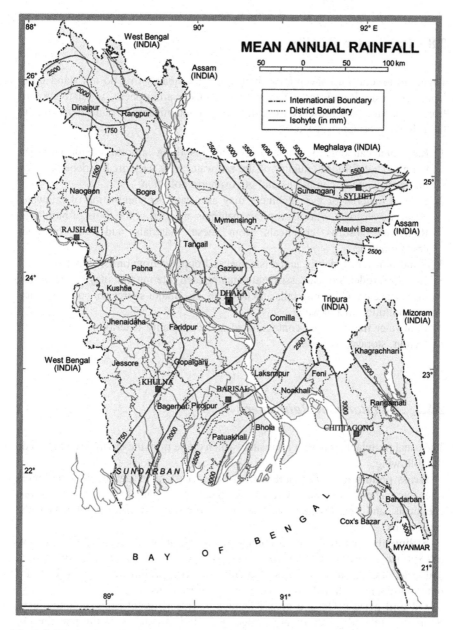

Fig. 2.4 Distribution of rainfall in Bangladesh. (P) *Source* GoB, Office of the Prime Minister, Bangladesh [43]

can be made available per m^2 of catchment area each year for development of a rainwater-based water-management system. In many countries in the world receiving half the rainfall of Bangladesh practice rainwater harvesting for their water-resource management.

2.8 Rainwater Quality

There are several misconceptions about rainwater quality that do not stand up to laboratory tests. Rainwater, when produced in the atmosphere, is relatively free from impurities. The presence of pollutants or contaminants in rainwater is due to the following established facts:

1. Rainwater can only be contaminated by absorbing airborne chemicals while falling. Therefore, the quality of the rainwater may vary greatly depending on the local climatic conditions and atmospheric pollution.
2. Surface material of the catchment may also cause contamination of rainwater when it overflows. Various roofing materials are used in the roofs of buildings. Rainwater absorbs chemicals from these materials and thus becomes contaminated.
3. Wind-blown dirt, debris, and dusts, faecal droppings from birds, animals, insects etc., dead rodents or carcasses and contaminated littering that may be present on the surface on which rain falls and flows are also major sources of contamination and pollution of rainwater.
4. Rainwater may also be contaminated and the quality deteriorated during collection, storage, and use.

2.8.1 Pollutants in Rainwater

The usual pollutants found in rainwater, which come from the environment, include the following:

Gross pollutants: These pollutants include litters, plant debris, floatable materials, etc. These pollutants virtually harbour other pollutants such as various chemicals and microorganisms that are carried by surface runoff from the catchments.

Sediments: Sediments are a common component of rainwater collected from any surface. Construction sites, roadways, parking lots, industries, areas of exposed soils, etc., are the major sources of sediments. These sediments act as vehicle for other pollutants such as heavy metals and hydrocarbons, etc.

Oil and grease: Oil and grease include a wide range of organic compounds. Sources of oil and grease on any surface include leaks and breaks in machines or appliances, kitchen wastes, waste oil, and cleaning and maintenance of machineries carried on the surface.

Nutrients: Nitrogen and phosphorous are often found in rainwater collected from any roof. The presence of these nutrients on roofs is due to deposition of particulate matters that are generated by automobile emissions, industrial processes, etc. The presence of these nutrients in harvested rainwater is beneficial for using in lawn irrigation.

2.8.2   Microorganisms in Rainwater

In rainwater, there is little chance of having any microorganisms, but airborne microorganisms have a significant contribution to the bacterial contamination in rainwater collected from roofs [44]. The presence of microorganisms is noticed mostly in stored rainwater. Collected and stored rainwater may contain a range of microorganisms, although most of them will be harmless. The major cause of microbial contamination is due to the faecal contamination of rainwater originating from bird and small animal droppings on the catchments from where rainwater is collected. Therefore, the presence of *Escherichia coli* (*E. coli*) bacteria are very common in stored rainwater. Other pathogens—such as *Cryptosporidium*, *Giardia*, *Campylobacter*, *Vibrio*, *Salmonella*, *Shigella* and *Pseudomonas* spp—have also been detected in rainwater [45]. In several cases of rainwater harvesting in Maldives, the presence of *toxoplasmosis* in stored rainwater has been reported, and has been noticed to be increasing, which seems to originate from cat droppings [45].

2.8.3   Chemical Contamination of Rainwater

The presence of dissolved minerals in rainwater is very low; as such, it has been observed to be relatively aggressive. Rainwater can dissolve heavy metals and other impurities from materials of the catchment, pipe, and storage tank. In most cases, chemical concentrations in harvested rainwater are found to be within acceptable limits; however, in some cases higher concentrations of zinc and lead have been reported [46]. This could be due to leaching from metallic roofs and storage tanks or from atmospheric pollution.

2.8.4   Qualitative Changes in Rainwater

Other than contamination and pollution of rainwater due to absorption of various substances and chemicals from the environment, various qualitative changes occur in the collected rainwater. The probable qualitative changes of collected rainwater are as follows.

Acidic rainwater: As rain falls from clouds and flows onto any surface on the Earth, rainwater absorbs atmospheric gases such as CO_2, SO_2, NO_2, etc. When dissolved in rainwater and oxidized by oxygen present in the air, these oxides results in acids: Carbon dioxide makes carbonic acid (H_2CO_3); sulphur dioxide produces sulphuric acid (H_2SO_4); and nitrogen dioxide becomes nitric acid (HNO_3). The more these gases are present in the air, particularly in industrial areas, the more the rainwater will be acidic in nature as well as taste.

Salt in rainwater: In harvested rainwater near coastal belts, the presence of salt in the stored rainwater may be noticed. Sometimes violent winds from offshore draw various kind of sea salt aerosol, which ultimately finds its way onto the catchments nearby. However, the occurrence of such violent winds is seasonal and brings comparatively little amount of salt. Although the concentration of salt would be low in rainwater, it may discourage the use of harvested rainwater as potable water [47].

pH of rainwater: Theoretically the pH value of rainwater varies between 4.5 and 6.5 [48]. This is usually found to increase slightly as it flows on the catchments and is stored in reservoirs made of concrete or having surface material containing cement. The chemical phenomenon in this regard is that the acidic rainwater, when in contact with alkaline fibrous cement, tends to become more alkaline, i.e., there is a reduction of the acidic nature of rainwater. This is considered a favorable condition in rainwater harvesting.

Odour in stored rainwater: The presence of mineral salts, organic matter coming from catchments, and sunlight penetration into the reservoir by any means promotes the growth of algae on the reservoir walls and in the stored rainwater. Over time these algae decompose under anaerobic conditions (i.e., the absence of air), thus giving the water an odour resembling the smell of rotten eggs. Another source of odour is accumulated sediment at the bottom of the tank. Sediment-related odours are caused by the anaerobic fermentation within the sediment. Other odour-related problems may be caused due to collecting rainwater from asphalt or tar-based roof coverings or a plastic-based roofing surface, which can give off odours similar to those of tar, rubber, or plastic. Odours caused by dead animals, rodents (mice, cockroaches, rats, etc.) and insects are altogether different from those caused by anaerobic fermentation. Such odours indicate the presence of drowned carcasses in the reservoir.

Softness: Water becomes "hard" due to dissolved calcium or magnesium ions, which are absent in rainwater, thereby making it soft in nature; this water leaves no lime scale. Washing clothes and hair in rainwater requires less detergent. The use of rainwater eliminates the need for a water softener and produces less-polluted wastewater. The presence of fewer minerals in rainwater also saves plumbing fixtures from wear and tear.

References

1. WW (2010). University of Illinois 'Cloud Types'. http://ww2010.atmos.uiuc.edu/(Gh)/guides/mtr/cld/cldtyp/home.rxml Retrieved on 14 Jan 2016
2. Pidwirney M (2016) 'Understanding physical geography' Chapter 8, thunderstorms, mid-latitude cyclones and hurricanes. Our Planet Earth Publishing, British Columbia, Canada, p 4
3. Weatherchatplus (2006) 'Rain' Weather Chat Forum. Climate Changing Centre. http://s2.excoboard.com/Weatherchatplus/124953/789046. Retrieved on 6 Aug 2015
4. Weatherchatplus (2006) 'Rain' Weather Chat Forum. Climate Changing Centre. http://s2.excoboard.com/Weatherchatplus/124953/789046. Retrieved on 14 Jan 2016
5. Elert G (2001) 'Diameter of a raindrop' The Physics Factbook, taken from "characteristics of particles and particle dispersoids. Handbook of Chemistry and Physics, 62nd edn. CRC, New York, 1981. http://hypertextbook.com/facts/2001/IgorVolynets.shtml. Retrieved on 06 Aug 2015
6. Henkel M (2015) 21st century homestead: sustainable agriculture II: farming and natural resources. Lulu.com, p 321
7. Ministry of Earth Science, Southeast Monsoon (2010) End of season report. India Meteorological department. http://www.tropmet.res.in/~kolli/MOL/Monsoon/year2010/Monsoon-2010.pdf. Retrieved on 14 Jan 2016
8. Facts About India Forum (2015) Distribution of rainfall in India. http://www.facts-about-india.com/distribution-of-rainfall-in-india.php. Retrieved on 12 Aug 2015
9. Food and Agriculture Organization (FAO) (1989) Irrigation water management: irrigation scheduling', training manual no. 4, Annex II: rainfall measurement*. http://www.fao.org/docrep/t7202e/t7202e09.htm Retrieved on 12 Aug 2015
10. Explore South Africa (2015) South Africa weather and climate. http://www.sa-venues.com/no/weather.htm. Retrieved on 12 Aug 2015
11. International Livestock Research Institute (ILRI) (2015) 'Rainfall and rainfall variability in Africa' posted by, Philip Thornton in ILRI News. http://news.ilri.org/2014/07/23/new-map-average-annual-rainfall-in-africa/. Retrieved on 12 Aug 2015
12. University of Nebrasca, Omaha (20000, 'North America', Physical Geography. http://maps.unomaha.edu/Peterson/geog1000/Notes/Notes_Exam4/NAmerica.html. Retrieved on 13 Aug 2015
13. Natural history on the net (2015) 'South America', climate. http://www.naturalhistoryonthenet.com/Continents/south_america.htm. Retrieved on 13 Aug 2015
14. Keen C (1997) 'Greatest places physical geography', Amazon Basin. http://www.greatestplaces.org/notes/amazon.htm. Retrieved on 13 Aug 2015
15. Maps of World 'South America Climate'. http://www.mapsofworld.com/south-america/geography/south-america-climate.html Retrieved on 13 Aug 2015
16. Extreme Science (2015) Driest Place: Atacama Desert, Chile. http://www.extremescience.com/driest.htm. Retrieved on 16 Aug 2015
17. Maker Media (2013) In Peru, Engineers Make Water out of Thin Air. http://makezine.com/2013/03/07/in-peru-engineers-make-water-out-of-thin-air/. Retrieved on 16 Aug 2015
18. International Centre for Terrestrial Antarctic Research (ICTAR) "Environment, Antarctica" http://www.ictar.aq/environment7.cfm?object_id=16&page_obj_id=206. Retrieved on 16 Aug 2015
19. Wikipedia, Climate of Antarctica' Precipitation. http://en.wikipedia.org/wiki/Climate_of_Antarctica. Retrieved on 16 Aug 2015
20. Australian Bureau of Statics (ABS) (2013). Year Book Australia, 2012 Australia's Climate. http://www.abs.gov.au/ausstats/abs@.nsf/Lookup/by%20Subject/1301.0~2012~Main%20Features~Australia's%20climate~143. Retrieved on 16 Aug 2015
21. Bureau of Materiology, Government of Australia (2015) Annual climate statement 2014. http://www.bom.gov.au/climate/current/annual/aus/. Retrieved on 16 Aug 2015

22. National Water Commission Government of Australia (2007) 'Rainfall distribution' Australian Water Resource 2005. http://www.water.gov.au/WaterAvailability/ Whatisourtotalwaterresource/Rainfalldistribution/index.aspx?Menu=Level1_3_1_2. Retrieved on 17 Aug 2015

23. Babkin VI 'Atmospheric precipitation of the earth' Hydrological cycle Volume-II. http:// www.eolss.net/Sample-Chapters/C07/E2-02-03-01.pdf. Retrieved on 17 Aug 2015

24. Schewe PF (2010) 'Rainy Spells Extended For Europe' INSIDE SCIENCE NEWS SERVICE. http://www.insidescience.org/content/rainy-spells-extended-europe/1400. Retrieved on 17 Aug 2015

25. SJS Wiki (2012) Earth Science 7-Southeast Asia2, precipitation. http://wiki.sjs.org/wiki/ index.php/Earth_Science_7-Southeast_Asia2. Retrieved on 17 Aug 2015

26. Phys.org (2015) Climate change and the South Asian summer monsoon. http://phys.org/news/ 2012-06-climate-south-asian-summer-monsoon.html. Retrieved on 17 Aug 2015

27. Abo-Hassan AA (1981) 'Rangeland Management in Saudi Arabia'. https://journals.uair. arizona.edu/index.php/rangelands/article/viewFile/11654/10927. Retrieved on 20 Aug 2015

28. Al-Jaloudy MAS (2001) Country pasture/forage resource profiles, JORDAN. http://www.fao. org/ag/agp/agpc/doc/counprof/jordan/jordan.html. Retrieved on 20 Aug 2015

29. Nepesov MA, Vitkovskaya TV, Kirsta BT (1999) 'Precipitation Use' in desert problems and desertification in Central Asia. The Researchers of the Desert Institute, Springer.

30. India Today 'Five wettest places in India you should know about'. http://indiatoday.intoday. in/education/story/five-wettest-places-in-india/1/445735.html. Retrieved on 30 Aug 2015

31. Maps of India Rajasthan 'Weather'. http://www.mapsofindia.com/maps/rajasthan/geography-history/weather.html. Retrieved on 30 Aug 2015

32. Meteo.co.in 'Rainfall and Monsoon statistics'. http://www.meteo.co.in/monsoon2.cfm. Retrieved on 30 Aug 2015

33. Facts-about-india.com 'Distribution of Rainfall in India'. http://www.facts-about-india.com/ distribution-of-rainfall-in-india.php. Retrieved on 31 Aug 2015

34. Banglapedia (2006), cited in 'Impact of Climate Change on Heavy Rainfall in Bangladesh', Final Report (2014). Institute of Water and Flood Management (IWFM)

35. MOU Tours and Travels, 'Land and Climate. http://www.mot-bd.com/index.php/bangladesh/ land-and-climate. Retrieved on 31 Aug 2015

36. World Bank (2000) Bangladesh: climate change and sustain-able development. Rural Development Unit, South Asia Region, World Bank, Dhaka

37. Shahid S, Khairulmaini OS (2009) 'Spatio-temporal variability of rainfall over Bangladesh during the time period 1969–2003'. Asia Pac J Atmos Sci 45(3):386. http://www.scribd.com/ doc/36927536/Rainfall-over-Bangladesh#scribd. Retrieved on 18 Jan 2016

38. Sharmeen F, Mafizur Rahman Md (2011) Characterizing rainfall trend in Bangladesh by temporal statistics analysis. In: 4th Annual Paper meet and 1st civil engineering congress, Dhaka, Bangladesh. http://www.iebconferences.info/359.pdf. Retrieved on 1 Sep 2015

39. International Business Publication (2012). 'Bangladesh country study guide' volume 1 strategic information and guide. IBP, USA, p 58

40. Institute of Water and Flood Management (IWFM) (2014) 'Impact of climate change on heavy rainfall in Bangladesh', final report. http://teacher.buet.ac.bd/akmsaifulislam/reports/ Heavy_Rainfall_report.pdf. Retrieved on 18 Jan 2016

41. The World Bank Data, 'Average precipitation in depth (mm per year)'. http://data.worldbank. org/indicator/AG.LND.PRCP.MM. Retrieved on 1 Sep 2015

42. Food and Agriculture Organization, electronic files and web site. http://www.indexmundi. com/facts/bangladesh/average-precipitation-in-depth

43. Government of the Peoples Republic of Bangladesh 'Maps of Bangladesh' Mean Annual Rainfall of Bangladesh, Prime Minister's Office Library, Dhaka. http://lib.pmo.gov.bd/maps/ images/bangladesh/Rainfall.gif. Retrieved on 18 Jan 2016

44. Evans CA, Coombes PJ, Dunstan RH (2006) 'Wind, rain and bacteria: the effect of weather on the microbial composition of roof-harvested rainwater'. Water Res 40(1):37–44. Epub 2005, Dec 15

45. Namrata Pathak, and Han Heijnen, 'Rainwater Harvesting and Health Aspects Working on WHO guidance'. http://www.ctahr.hawaii.edu/hawaiirain/Library/papers/Pathak_Namrata.pdf. Retrieved on 1 Sep 2015
46. World Health Organization (WHO). 'Rainwater harvesting', Chapter 6'. http://www.who.int/water_sanitation_health/gdwqrevision/rainwater.pdf. Retrieved on 2 Sep 2015
47. Eautarcie, Sustainable Water Management for the world 'Rainwater Quality in a Cistern'. http://www.eautarcie.org/en/03b.html. Retrieved on 2 Sep 2015
48. Carlos Schmidt Quadros (2010). 'Rainwater Harvesting' Case Study: FCT/UNL Campus, University NOVA, Lisboa. http://run.unl.pt/bitstream/10362/4799/1/Quadros_2010.pdf. Retrieved on 21 Sep 2015

Chapter 3
Rainwater-Harvesting Technology

Abstract Rainwater harvesting in buildings involves technology for its proper planning, design, installation, operation, and maintenance. Two major scopes of rainwater harvesting are (1) the use of rainwater for all general purposes and (2) recharging groundwater. In both cases, various functional techniques must be applied. The extent of technological involvement depends primarily on the introduction of various conditioning or treatment technologies in the harvesting system: sedimentation, filtration, and disinfection. Based on the application of these conditioning or treatment technologies, the rainwater-harvesting system in buildings may be classified as (1) a direct-use system, (2) a nonfiltered system, (3) a filtered system, and (4) a complete system. The development process of a rainwater-harvesting system in buildings has four aspects: (1) planning, (2) design, (3) construction, and (4) maintenance. In all aspects of rainwater harvesting–system development, systematic approaches must be followed to ensure its sustainability. In this chapter, the major scopes of rainwater harvesting are introduced. The extent of technological involvement in the harvesting system is delineated. All aspects of rainwater-harvesting development in buildings and their systematical approaches are discussed. Finally, the problems and prospects of rainwater harvesting in buildings are identified.

3.1 Introduction

Rainwater harvesting is a subject of science and involves technology for its design, installation, operation, and maintenance. For rainwater harvesting in buildings, a set of jobs must be performed to attain the objectives. Rainwater harvesting in a building deals with the management of rainwater and is based on hydrological philosophy as well as sanitation and plumbing technology. In this chapter, the technical jobs that must be performed to harvest rainwater in buildings are discussed exclusively.

© Springer International Publishing Switzerland 2017
S.A. Haq, PEng, *Harvesting Rainwater from Buildings*,
DOI 10.1007/978-3-319-46362-9_3

3.2   Scopes of Rainwater Harvesting

Rainwater harvesting may be defined as the use of rainwater before it joins the water courses from where its collection is either limited or costly. Rainwater harvesting in buildings implies the collection of rainwater that fall on a building or its premises as well as the use of the collected rainwater. In this system, rainwater is intercepted, diverted, stored for future use, and distributed, and a portion is infiltrated into the ground before draining the excess rainfall. Therefore, in a broader perspective, rainwater-harvesting technology is applied in two major sectors as mentioned below.

1. General-purposes use
2. Recharging groundwater.

3.2.1   Rainwater for General-Purposes Use

In buildings of different occupancies, there is a need of water for variety of uses starting from drinking to cutting metals. Whatever may be the purpose of use, it is well assured that in all purposes of water use, rainwater can be used for the same purpose subject to the proper treatment of the rainwater needed for that purpose. General-purposes use of rainwater in a building involves wide range of functional techniques, which are discussed later in the text.

3.2.2   Rainwater for Groundwater Recharging

When buildings must be built in an area that experiences acute shortages of water—particularly in areas where the source of water is predominantly groundwater—over-extraction of groundwater might cause various hydro-geological problems. Rapid lowering of the groundwater level and its depletion are major concerns. In Dhaka city of Bangladesh, the average lowering rate of groundwater is >3 m/y [1]. In such a situation, groundwater recharging is the most effective approach for improving the groundwater situation, and the use of rainwater would be an appropriate choice for this purpose.

3.3   Functional Techniques in Rainwater Harvesting

For harvesting rainwater, a series of jobs must be performed. In each job, the appropriate technique must be applied; otherwise the objective of harvesting might not be achieved. The techniques involved in rainwater harvesting for general-use purposes differ from those involved in groundwater recharging. The functional techniques involved in harvesting general-use rainwater and groundwater recharge are listed below.

3.3.1 Functional Techniques for General-Use

The incorporation of a rainwater-harvesting system in a building requires various natures of functions to be performed that involve technology for its effective and efficient performance. The following techniques must be adopted for the installation of a general-use rainwater-harvesting system in a building.

1. Collecting rainwater from suitable catchments
2. Flushing out the first rainwater
3. Storing rainwater
4. Qualifying rainwater according to the purpose of use
5. Supplying rainwater to the points of use
6. Draining excess rain water.

All of these techniques are discussed elaborately in subsequent chapters. A schematic diagram, showing all major components of rainwater harvesting in a building for its general-purpose uses, is shown in Fig. 3.1.

3.3.2 Functional Techniques for Groundwater Recharge

In recharging groundwater using rainwater, various functional techniques are involved as follows:

1. Collecting rainwater from suitable catchments
2. Conveying rainwater to the recharge structure
3. Qualifying rainwater according to the method of recharging adopted
4. Draining of excess rain water.

3.4 Extent of Technological Involvement

The technological involvement in all of the components of rainwater harvesting in buildings mentioned herein must be properly enacted in all cases and situations. It should be noted that among all of the components, the employment of some components are essential, and the rest are optional. Rainwater harvesting in buildings can be performed in a very simple way, as shown in the Fig. 3.2, employing very simple technology. It may also be developed in other ways

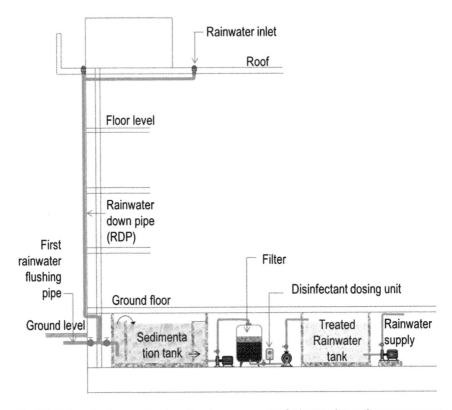

Fig. 3.1 Schematic diagram showing all major components of rainwater-harvesting system meant for general use in a building

employing various sophisticated and automated technologies in almost all of the components of rainwater harvesting depending on the demand and desire of the users of the building. Various ways of developing a rainwater-harvesting system in a building by combining different essential and optional components of a harvesting system are discussed below.

3.4.1 Direct-Use System

In this methodology, rainwater is collected directly from the catchments and subsequently used. First, rainwater flushing can be employed or may be avoided depending on the purpose of the rainwater use. This is the simplest and most basic form of rainwater harvesting. In this system, only a catchment area and storage tank are needed. A screen is provided on the rainwater inlet or gutters as shown in Fig. 3.2. The catchment must be kept clean, and the storage tank must be maintained regularly.

Fig. 3.2 Schematic diagram
of a direct-use system of
rainwater harvesting in
buildings

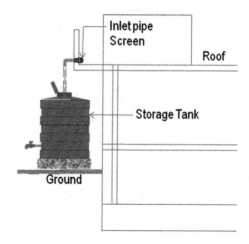

3.4.2   Nonfiltered System

In this methodology, sedimentation is facilitated in the storage tank by incorporating the necessary elements in the storage tank for the proper settling of suspended particles present in the rainwater. In this system, no extra settling tank is incorporated; rather, the storage tank is designed in such a way as to play dual role of settling and storage as shown in Fig. 3.3.

Fig. 3.3 Schematic diagram
of nonfiltered
rainwater-harvesting system
in buildings

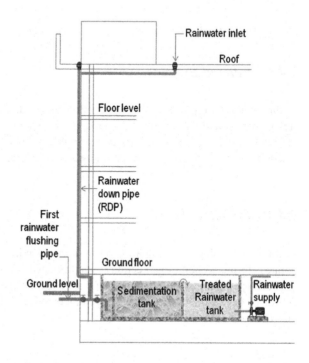

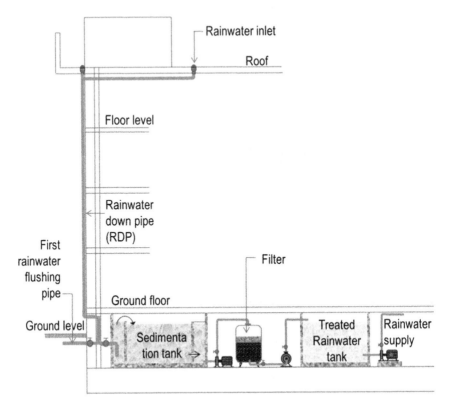

Fig. 3.4 Schematic diagram of a filtered rainwater-harvesting system in buildings

3.4.3 Filtered System

In this methodology, a filtration system is employed, which is installed after the sedimentation tank. Here an extra tank, as shown in the Fig. 3.4, is needed to collect the filtered rainwater from where the rainwater is supplied to various locations in the buildings where the rainwater is needed.

3.4.4 Complete System

In this methodology, all conditioning systems of rainwater harvesting are incorporated. In this system, disinfection is incorporated. This complete system should be developed for consuming rainwater as potable water. A schematic diagram shown in the Fig. 3.1 representing a complete system of rainwater harvesting for general-use purposes.

3.5   Aspects of Rainwater-Harvesting Technology

Rainwater-harvesting technology, either for general-use purposes in a building or for groundwater recharging, involves a wide range of approaches and functional techniques for its planning, design, construction, and maintenance. The success and efficiency of rainwater harvesting in buildings fully depends on these four aspects of technology. Ignorance and/or failure of any of these aspects might result in abandonment of establishing this system in a building.

3.5.1   Planning Aspects of a Rainwater-Harvesting System

A reinforced concrete building is supposed to have a maximum useful structural life of 75 years [2]. In this long span of life, the environmental scenario, particularly the water-environment scenario, of the area might be changed, for which alternative options for water management might be required for the building. After construction of a building, the incorporation of a rainwater-harvesting system, as a supplemental water-management system, will be difficult, if not impossible, particularly in terms of accommodating the storage and conditioning structures of appropriate size at the proper locations. Therefore, while preparing the architectural drawing of a building, planning for the incorporation of rainwater-harvesting elements must be performed in light of following planning perspectives.

1. Synchronizing the harvesting system with the normal water-supply and—drainage system
2. Positioning the components of the rainwater-harvesting system
3. Incorporating the provisions for maximizing collection in case of shortage.

To develop a well-planned rainwater-harvesting system in a building, the following jobs must be performed during the planning approach.

1. Identify the purposes requiring the use of rainwater
2. Plan the catchments
3. Plan the collecting system
4. Plan the manner of storing and assign the storage system a location
5. Plan the conditioning system
6. Plan the supplying system
7. Plan the location of the recharge structure.

3.5.2   Design Aspects of a Rainwater-Harvesting System

The design aspect of rainwater harvesting involves optimum sizing of the components of the harvesting system. The objective of designing the components is to ensure the following:

1. Run the system efficiently
2. Keep the system safe from pollution or contamination
3. Save energy.

The jobs in designing the system include the following:

1. Estimate the amount of rainwater to be collected and stored
2. Size the catchment area needed
3. Design the collection and drainage pipes and its appurtenances
4. Design the pumps
5. Design the tanks, e.g., storage tanks, sedimentation tanks, etc.
6. Design the filter units
7. Design the recharge structures.

3.5.3   Construction Aspects of a Rainwater-Harvesting System

The construction aspect of a rainwater-harvesting system focuses on achieving the following objectives:

1. Durability
2. Safety
3. Stability.

The jobs to be performed while developing a rainwater-harvesting system in a building include the following:

1. Earth work for underground construction of storage tanks, laying pipes, etc.
2. Collect drain pipe, fittings and installation
3. Supply pipe fitting and installation
4. Install tanks
5. Install pumps and accessories
6. Construct or install treatment or conditioning system.

3.5.4   Maintenance Aspects of a Rainwater-Harvesting System

To ensure sustainability of the rainwater-harvesting system the following aspects of maintenance of the system must be achieved.

1. Acceptable hygienic conditions
2. Smooth functioning of the system
3. Longer life.

With a view to achieving the above-mentioned objectives, maintenance of the following components of the rainwater-harvesting system must be maintained at regular intervals.

1. The catchments
2. All pipes and accessories including its inlets
3. Gutters or channels conveying rainwater
4. Storage tanks
5. Filtering units
6. Pumping units.

3.6 Planning Approach for a Rainwater-Harvesting System

3.6.1 Identifying the Purpose of Use

At the onset of planning of the development of any building, local building-development rules or the regulations, codes, and permit system must be envisaged regarding the obligations of installing a rainwater-harvesting system for that particular area where a building is supposed to be developed. It should also be known whether harvesting should be performed for general-purpose use, for groundwater recharging, or for both. The availability of applicable incentives must be sought. The property or building owner and the concerned stakeholders must be consulted regarding the objectives, constraints, and cost of the system to be incorporated for rainwater harvesting. Accordingly, all necessary provisions must be incorporated into the architectural planning of the building.

3.6.2 Planning the Catchments

After determining the amount of rainwater to be used for various purposes of harvesting rainwater, the catchments and the area of catchments to be used should be determined. For effective use of the catchments, these areas should be kept free from other uses. Finishing material of these catchments must be chosen considering the possibility of contaminating the rainwater to be collected.

3.6.3 Planning the Collection

After determining the volume of rainwater to be harvested, the nest step is to calculate the amount of rainwater that can be collected from the catchments. If water available from the roof is found to be satisfactory, then the problem is solved. If available water from the roof is found to be less than the required amount, then other catchments—such as a porch, a vertical wall, a veranda or a balcony—must be considered. When all of the potential catchments fail to provide the required rainwater, then various methods of maximizing the collection of rainwater should be tried; these methods are discussed elaborately in the chapter on rainwater collection.

3.6.4 Planning the Storage System

In planning the storage of rainwater, various aspects of storage should be considered. The first consideration is the size of storage, and the next important consideration is the location where it should be placed. Choosing the suitable location for storage again depends on various factors that are elaborately discussed later in an chapter devoted exclusively to storage.

3.6.5 Planning the Conditioning System

While planning for the conditioning system, first of all the applicable conditioning-system components are to be chosen depending on the quality and type of use of the rainwater. Planning is also required to place the components of conditioning elements at the proper locations. In placing, the location of the storage tanks plays very important role.

3.7 Design Approach of a Rainwater-Harvesting System

3.7.1 Estimating the Amount of Rainwater

After making the decision to install a rainwater-harvesting system, the purposes of use should be identified, which will dictate an estimate of the amount of rainwater to be collected. For general-purpose uses of rainwater, the purpose for which the rainwater should be used should be determined. Then it is necessary to determine the amount of water available from various other sources. The amount of water that falls short will be fulfilled by the rainwater. These estimations are needed basically to determine the volume of storage tanks required for the different uses of rainwater.

3.7.2 Sizing the Catchment Area

After estimating the amount of rainwater to be used, the required catchment area should be calculated based on the rainfall data available. If the amount of rainwater to be used is relatively small, the roof might be sufficient to generate that volume. If roof becomes insufficient, then other potential catchments of the building can be chosen. When all of the potential catchments of a building cannot fulfill the rainwater demand, then available catchments outside the building can also be considered.

3.7.3 Designing of Collection and Overflow Pipes

It is preferred to have one collection pipe receiving rainwater from multiple inlets of one catchment or multiple catchments at different locations. All of the inlets are finally connected to the collection pipe through branch pipes leading to inlets. Summation of the tributary areas of connected inlets will be the total catchment area for the collection pipe for which it must be designed. The size of overflow pipe should be at least one size larger than the diameter of the collection pipe.

3.7.4 Designing Other Components

Designing other components of a rainwater-harvesting system—such as a pump, storage tanks, sedimentation tanks, filter units, and recharge structures—can be performed individually. The design procedures are discussed in subsequent relevant chapters.

3.8 Construction Approach for a Rainwater-Harvesting System

It should never be thought that a rainwater-harvesting can be developed after the construction of the building structure, although it can be if necessary. If for any reason the system is to be developed after construction of the building structure, the system may not be well-integrated with the existing plumbing system of the building. It will be good practice to construct the rainwater-harvesting system along with the construction of a building's structure and its services. The below-grade RCC or masonry reservoirs within the building should be constructed during construction of the structural components at that level. The piping and pumping units of a rainwater-harvesting system must be installed along with the piping and pumping units of plumbing installations, respectively. If the below-grade reservoirs must be built outside the building, then the rainwater-harvesting can be constructed at a suitable time during development of plumbing system.

3.9   Maintenance Approach to a Rainwater-Harvesting System

Every component of a rainwater-harvesting system must be regularly maintained at certain intervals. The frequency of maintenance requirements of the major components of a rainwater-harvesting system are suggested below.

Catchments: The areas from which the rainwater will be collected must be swept every month to remove leaves, litter, and dirt. Catchments must be washed off with water when dust or dirt accumulates, thus diverting runoff away from the filter or storage tank before rain starts. Tree branches hanging over the catchments must be trimmed when required.

Gutters: Gutters must be kept clean by washing out bird droppings, leaves, etc., with water before it rains or after heavy winds. The cleanliness, stability, alignment, and slope of the gutters must be checked every 3 months as well as after a storm [3] to ensure their proper functioning. After a heavy shower, the slope of gutter should be checked specially and corrected if any deviation is observed.

Pipes: All of the pipes—including inlet screens installed for rainwater harvesting—must be maintained and monitored in the way gutters should be addressed. The pipes must be checked regularly for leakages and must be repaired when required.

First-flush device: The first-flush device must be checked for cleanliness. It should be cleaned before and after the monsoon rain starts and after every rooftop-cleaning operation.

Treatment system: Filters should be cleaned at least once in a year before the monsoon rain begins and backwashed after the monsoon rain is over. A chlorine level at or slightly greater than 0.5 mg/L and a pH level of 6.5–8.5 must be ensured by testing weekly and after heavy rains [3]. The UV light must be inspected weekly to keep it free from scum.

Storage tank: Extra care for the storage tank should be taken in the regular maintenance program of the rainwater-harvesting system. The tank must be cleaned before the monsoon rain starts. Occasionally the tank must be checked for the development of any cracks or leakages. If any crack or leakage is noticed, it must be repaired at the earliest possible time. In case of underground storage tanks, the growth of trees on or nearby the tank must be checked all of the time, and roots that might affect the tank must be cut when necessary. The lid of the tank must be ensured for its sturdiness, and there must be no gap between lid and its rim. At the end of the overflow pipe, the insect screen must be well secured at all times.

Equipment and accessories: All of the equipment and accessories installed in the rainwater-harvesting system, e.g., pump, valves, etc., must be checked at least once a year.

Water quality: The presence of mosquito larvae in stored rainwater must be checked for every 3 months [3]. Accumulation of the sludge level must be monitored with the same frequency if rainwater is to be used for drinking purposes. An E. coli test should be performed at the initial development stage of the harvesting system to identify the risk of bacteriological contamination of the stored rainwater. In addition, this test should be performed when the system or a part of the system is altered.

3.10   Prospects of Rainwater Harvesting

It is believed that rainwater is harvested for mitigating the crisis of water prevailing in particular areas. Virtually this is the major and direct implication of rainwater harvesting, but there are various other indirect implications of rainwater-harvesting systems being installed in buildings. The prospective implications of rainwater harvesting in buildings are as follows:

1. Decreases the pressure of using surface water and groundwater
2. Decreases the cost of water consumed
3. Increases groundwater availability
4. Decreases urban flooding and water logging
5. Harvested rainwater saves energy. A 1 m increase in groundwater level saves 0.40 kWh of electricity in extracting groundwater [4]
6. Aids the development of energy-saving "green" buildings
7. Helps to provide water during any crisis
8. Intervention in adaptation and decrease in vulnerability due to climate change
9. Renewable resource that does not pose any negative impacts on the environment
10. Salt-free source of water and its use in groundwater recharging decreases salt accumulation in soil.

3.11   Problems of a Rainwater-Harvesting System

It is not that rainwater harvesting always brings prospects in the water-management sector. It also has some negative implications in built environments. Following are those negative implications of rainwater harvesting in building development on surrounding environment.

1. The availability of rainwater may be limited due to occurrence of long dry spells.
2. Rainwater quality may not be consistent due to variability in air pollution and other sources of contamination.
3. Regular consumption of mineral free rainwater may cause nutritional deficiencies.
4. The initial investment cost of the harvesting components is relatively high.
5. Lack of proper operation and regular maintenance will restrict the desired output from the system.

6. Storage space may hinder the installation of other building service elements at suitable locations.
7. There is extra cost for a separate plumbing system conveying the rainwater.
8. Rainwater harvesting under compulsion may not be agreeable to the owner or developer.
9. The over-recharging of groundwater may create problems with underground structures.

References

1. Sos-arsenic.net (2015) Ground water. http://sos-arsenic.net/english/groundwater/index.html. Retrieved on 18 Jan 2016
2. Flager FL (2003) The design of building structures for improved life-cycle performance. Masters thesis. Department of Civil and Environmental Engineering, Massachusetts Institute of Technology, p 13
3. Victorian Government (2009) Making sure your private water supply is safe, Rainwater, Food Safety and Regulatory Activities, Department of Human Services, Melbourne, Australia. https://www2.health.vic.gov.au
4. Government of the Republic of Maldives (2009) Guidelines and manual for rain water harvesting in Maldives. Technical Support of World Health Organization, Male, Maldives, p 25. http://www.searo.who.int/maldives/documents/Maldives__Guidelines_and_manual_of_ Rain_Water_Harvesting_in_Maldives_2009.pdf?ua=1. Retrieved on 09 Sep 2015

Chapter 4
Water Requirement

Abstract A building in use may require water either related to the purpose of the building or for the user of the building. In addition, water is also required to keep the building usable. The use of water in a building may be for a single unique purpose or multiple purposes. To harvest rainwater in a building, it is of the utmost necessity to identify all of the purposes for the use of water and then to estimate the water requirement for those purposes. Water use in a building is mostly a continuous phenomenon throughout the day, and in some cases it is periodical or in few cases occasional. Considering the usual pattern and major purposes for water to be used in a building, the daily water requirement for the purposes of using water in a building should be determined. Water requirement for particular purposes varies depending on multiple factors such as behavioral, economical, social, cultural, environmental, and probable hazards. Therefore, the water requirement for a particular single purpose is not identical or unique for buildings of various occupancy types at various locations. Water requirement, and thus the demand for water for a particular purpose, varies from person to person, from time to time, and from place to place. As a result, the water requirement in a building for a particular purpose and occupancy type varies considerably for various locations. In this chapter, various water requirements for persons and thus for buildings are mostly identified and mentioned in tabular forms. In most cases, the water requirement or demand is furnished on a daily basis. In a few cases, the requirement is given in the context of particular considerations. The water requirement for some purposes, in particular areas on the globe, is also furnished.

4.1 Introduction

A water-supply system in a building is developed because there might be demand for it for various purposes at various locations of the building, but the source of water may be elsewhere. The demand for water implies adequate water quantity and acceptable quality needed for various purposes. At the onset of planning for a water-supply system, the actual requirement of water is determined. In buildings of

© Springer International Publishing Switzerland 2017
S.A. Haq, PEng, *Harvesting Rainwater from Buildings*,
DOI 10.1007/978-3-319-46362-9_4

different occupancies, water is demanded in varied quantities for various purposes. The requirement of water for those purposes may also differ in nature. In this chapter, the water requirement for various purposes is discussed. When the requirement of water for all of the purposes in a building is determined; then its availability can be ensured. If the supply can be ensured from a single source, either from a water main, an independent tube well, or a combination thereof, no other alternative source will be required. When those sources are found not dependable, either fully or partially, to meet the total demand of water, then rainwater, if found reliable as the next option, is considered. Generally rainwater is used as a supplementary source of water. Therefore, total water requirement in a building should be determined first. Then the quantity and quality of water available from the main or any independent source is determined and judged. If the availability and usability of water from these two sources become limited, then the rest of the water demand may be met by rainwater. Therefore, to determine the quantity of rainwater requirement in a building, the total amount of water needed for the building should be known first. While planning and designing the water-supply system in any building, it is necessary to first-identify all of the purposes for water use in the building.

Many factors are involved in estimating the per-capita requirement for water in a building. Therefore, it is not possible to determine accurately the actual demand of water in a building; however, judiciously considering those varying factors of per-capita water demand, the water requirement in the building could be properly estimated. In this chapter, the water requirement for various purposes in various characters of occupancies of buildings is furnished to help in estimating the total water requirement in a building.

4.2 Various Types of Water Demand

In building, water is not just used by the users for their own purposes. There are various other uses of water in buildings such as safety, sanitation, and recreational purposes. Following are the various sectors in a building and the premise where rainwater can meet full or partial water requirements.

1. Occupants of buildings of various occupancies
2. Building construction, maintenance, and demolition
3. Visitors
4. Swimming pools, spas, water bodies
5. Fire suppression
6. Road, pavement washing
7. Irrigation and gardening
8. Poultry or animal rearing

9. Fountains, waterfalls, water cascades, etc.
10. Air conditioning and heating system
11. Cleaning building sewers
12. Special purposes

4.3   Domestic Water Demand

The basic quantity of water required in a residential or any dwelling unit is predominantly for drinking, cooking, personal hygiene and sanitation, etc., which is termed the "domestic water demand." The demand of such water varies widely, which mainly depends on the habits, social status, climatic conditions, religion, and customs of the people living in the building. According to Gleick (1996), the basic quantity of water required for a person is 50 L [1, 2]. Per IS 1172–1963, under normal conditions, the domestic consumption of water in India is approximately 135 litres/capita/day [3]. In Europe, Romanian people consume the highest amount of water, i.e., approximately 294 litres/day, and the people of Estonia consumes the lowest, i.e., approximately 100 litres/day [4]. On average, people of the United Kingdom, Finland, France, and Luxemburg consume approximately 150 litres/capita/day [4]. In some developed countries, such as North-America and Japan, this figure is as high as 350 litres/capita/day [5] because of the use of various-water consuming amenities such as air coolers, air conditioners, lawn maintenance, automatic household appliances etc. In contrast in sub-Saharan Africa the daily per-capita consumption of water is 10–20 litres/day [5]. Details of various domestic-water consumption in India are listed in Tables 4.1 and 4.2 lists the domestic-water consumption in England and Wales.

Table 4.1 Details of various domestic-water consumption in India

Sl no	Purpose	Amount (lpcd)
1	Drinking	5
2	Cooking	5
3	Bathing	55
4	Flushing toilet	30
5	Clothes washing	20
6	Utensil washing	10
7	House washing	10
Total		135

Source [3]

Table 4.2 Details of various domestic-water consumption in England and Wales

Sl no	Purpose	Percentage (%)	Amount (lpcd)
1	Kitchen and other use	42	63
2	Baths and showers	22	33
3	Toilet flushing	34	51
4	External use	2	3
Total		100	150

Source UK Department of the Environment (1997) [6]

4.4   Institutional and Commercial Water Demand

Buildings of academic institutions (e.g., public and private universities, colleges, schools etc.), charitable institutions (e.g., hospital, clinics, health centers, railway and bus stations, etc.), commercial enterprises (e.g., office buildings, warehouses, stores, hotels, shopping centers, cinema houses, etc.), and religious institutions (e.g., mosques, temples, etc.) come under this category. The requirement for water in these varieties of institutional buildings in urban areas also varies considerably depending on the provided facilities that require water. Table 4.3 lists the water requirement in various institutional buildings having specific facilities.

4.4.1   Water Use in Office Buildings

The water-use pattern in office buildings is different compared with domestic-use patterns. In office buildings, flushing water closets and urinals consume the major portion of total water used. The average consumption of water in Jordanian office buildings is listed in Table 4.4

4.5   Water Demand for Common Utility Purposes

In a considerably large building complex, an additional quantity of water is required for general and common utility purposes such as washing and sprinkling on roads, cleaning building sewers, and watering parks, gardens, fountains, etc. These requirements might be fulfilled for abidance and meeting the requirement of building construction regulations. To meet the water demand for the areas and utilities of common use, additional water is stored or preserved somewhere in the building complex. The water requirements for these common areas and utilities are listed in Table 4.5.

Table 4.3 Water demand for various institutions and commercial enterprises [3]

Sl no	Type of institutional building	Litres per day (lpd)
1.	Hospital (including laundry)	
	(a) No. of beds exceeding 100	445
	(b) No. of beds not exceeding 100	340
2	Hostel	135
3	Nurses home and medical quarters	135
4	Restaurant (per seat)	70
5	(a) Factory where bathroom is provided	45
	(b) Factory without bathroom	30
6	Office	45
7	Hotel (per bed)	180
8	Cinema, concert hall, and theatre (per seat)	15
9	School	
	i. Day school	45
	ii. Boarding school	135
	iii. Nursery school	45
10	Sport ground	3.5 litres/m^2
11	Air-conditioning(general)	70 l/h/100m^2
12	Shopping center	180
13	Airport and sea port	70
14	Railway and bus station	
	a. Terminal	45–70
	b. Junction station for mail or express:	
	i. With bathing facility	70
	ii. Without bathing facility	45
	c. Intermediate stations where mail and express do not stop:	
	i. With bath	48
	ii. Without bath	23
15	Community center	180
16	Religious institution	180
17.	Shopping center	180
18	Group housing	180
19	Airport and sea port (international and domestic	70

4.6   Water Demand for Other Purposes

Water is needed for many more purposes and in many sectors other than those discussed so far. There are some special uses of water also that may be considered optional for some building complexes. The water requirement in these optional sectors shall be considered, if needed, while estimating the water demand in a building complex area or any building projects. The followings are those sectors.

1. Fire fighting
2. Recreational
3. Gardening and plantation
4. Animal rearing
5. Special uses

4.6.1 Water for Fire Fighting

Fire may develop in buildings due to short circuiting caused by faulty electric wires, the presence of flammable materials, explosions, criminal activity, or any other unforeseen mishap. If fires are not properly controlled and extinguished within the minimum possible time, they may lead to serious damage of lives and properties. Therefore, sufficient water is kept stored in individual buildings to fight against a probable occurrence of fire in the building.

Water for firefighting in buildings: In buildings, there is every chance of developing fire accidentally due to kitchen fire, electrical short circuit, and various combustible materials used in the building. To suppress such a fire, a huge amount of water is required. Rainwater can be used for fire fighting. To ensure water available during a fire, a considerable amount of rainwater must be kept stored in the building if there is no other source of water nearby, i.e., a street fire hydrant. The amount of rainwater to be stored for firefighting is to be determined from the

Table 4.4 Average water use in Jordanian office building

Sl no	Purpose	Percentage (%)	Amount (lpcd)
1	Water-closet flushing	63	28
2	Urinal flushing and bidets	3	1.5
3	Faucets (lavatory and kitchen)	17	7.5
4	Landscape	1	0.5
5	Cleaning	14	6.5
6	Others	2	1
Total		100	45

Source [7]

Table 4.5 Water requirement in various public utilities

Sl no	Purpose	Water requirements
1	Horticulture	4.5 million litres/acre/day
2	Sanitary sewer cleaning	3.0–5.0 litres/head/day
3	Road washing	5.0 litres/head/day

Source [3]

guidelines provided in the building or fire code applicable for the concerned areas. This water volume mostly depends on the fire-hazard level of the building. Most building codes categorize at least three levels of fire hazards: (1) "light hazard" includes residential homes, office buildings, educational institutions, etc.; (2) "ordinary hazard" includes factories or warehouse buildings; and (3) "high hazard" include buildings containing or manufacturing highly flammable materials. The quantity of water to be stored is usually measured in relation to pumping hours specified for the particular hazard level of the building. Ordinary-hazard buildings generally require 30 min to 1 h of pumping operation, and high-hazard buildings generally require 3–4 h of pumping operation. The required flow rate of water for firefighting may be as low as 1000 lpm [8] to as high as 9000 lpm [9].

Water for firefighting in building complexes: In large building complexes comprising various occupancies, a separate provision of water can be created for firefighting instead of storing water in every individual building for this purpose. In this case, the building complex is required to have sufficient water stored for fire-fighting, which can be met by rainwater if needed. In this system, a pressurized water-supply piping network is developed inside the building complex in which a sufficient number of fire hydrants are installed to draw water for fighting fire in nearby buildings. Generally fire hydrants are installed on the water mains at 100–150 m apart for covering multiple buildings to be served by nearby hydrant.

The quantity of water required for firefighting is generally estimated by using different empirical formulae. In India, the Kuiching's formula, as shown below, serves satisfactorily [10].

$$Q = 3182\sqrt{P} \tag{4.1}$$

where Q is a quantity of water required in litre/min, and P is the population of the building complex in thousands to be served.

The required number of fire streams having a flow of approximately 1136 lpm that can be operated simultaneously is determined by the following formula given by Kuiching [11].

$$F = 2.8\sqrt{P} \tag{4.2}$$

where F is the number of simultaneously operating fire streams, and P is the population in thousands to be served.

4.6.2   Recreational Water

Recreational-water structures—such as swimming pools, indoor fountains, water-falls, water cascades, etc.—are built for recreational purposes inside or outside a building. These water structures require a considerable amount of water. The water requirement for these recreational purposes depends on the size of the structure, particularly the size and flow rate through the nozzles used. For these recreational structures, water is not required to be drained and replaced in everyday. Depending on the maintenance and filtering system, once the structure is filled with water, that water can be recycled every couple of months. It is not wise to obtain water for such structures from a water main or a tube well. The use of rainwater for these purposes would be more justifiable in water-scarce areas. Table 4.6 lists information for estimation of the water requirement for some recreational purposes.

4.6.3   Water for Gardening and Plantation

A garden on the building premises is enjoyed by everyone. Building regulatory authorities permit to the construction of buildings in such a way as to keep considerable green space around where gardening or planting are the only activities allowed. However, the high cost of land does not always allow the owner or developer to fulfill such desires, which is then tried to be fulfilled alternatively by growing plants on the floor and the roof. Plantation and gardening in building grounds not only adds beauty but also help cool and improve the indoor environment.

Gardens or planters in buildings cannot be kept alive without sufficient sunlight and water. Sufficient water will be required for keeping plants, herbs, or trees alive and healthy depending on the size of garden or planter and the type of plants, herbs, or trees cultured. The water for gardening does not have to be pure or potable water but can wisely be provided by rainwater. Therefore, the amount of rainwater to be stored for this purpose must be estimated. Table 4.7 lists estimates of the water needed for horticulture inside or outside a building.

4.6.4   Water for Animal Rearing

In a building, some of the occupants might have various cattle or animal for rearing as hobby, security, or for the purpose of husbandry. For rearing these animals or cattle, water is needed, which must be included in the estimation of the water requirement in a building. Table 4.8 lists estimates of the amount of water needed for these purposes.

Table 4.6 Water requirement for recreational uses in a building

Sl no	Purpose	Water requirement
1	Fountain	3–5 lpm/nozzle
2	Swimming pool	4 % more than pool capacity

Source [3]

Table 4.7 Water requirement for horticulture in and around building [3]

Sl no	Purpose	Water requirements (litres/m²/day)
1	Kitchen garden	1.4
2	Sports grounds	3–5
3	Park and garden	2–3

Table 4.8 Water requirement for rearing animals, cattle, and poultry

Sl no	Animals/cattle	Water requirements (litres/capita/day)
1	Cow or buffalo	40–100 [3, 12]
2	Horse	40–50 [3]
3	Dog	8–12 [3]
4	Sheep or goat	5–10 [3]
5	Poultry	0.19 [12]

4.6.5   Water for Special Uses

Hot-water requirement: The use of hot water has diverse health and other benefits. Hot water helps in increasing the range of motion of the muscles. Body temperature and heart rate gradually increase, and the blood vessels dilate. As a result, the blood-circulation system improves. Good blood circulation helps bring nutrients and oxygen into the body; in contrast, it removes toxins and other unwanted substances that may otherwise restrict the blood flow. The top layers of the skin are cleansed well because the pores become enlarged and easily release moisture, which carries away chemical waste and dirt. Soaking any part of body in hot water helps dull pain experienced in that part of the body. Hot water also helps clean clothes, utensils, etc., particularly in removing oils from the utensils.

In most buildings, the hot water–supply system is developed generally for taking showers or bathing, dishwashing, personal cleansing, etc. These systems typically consist of a heating source that heats the water to a temperature of approximately 60 °C, which will kill most Legionella bacteria, which is then supplied directly or continuously circulated in the system. Side-by-side cold water is supplied to the mixing faucets of fixtures, where hot water is mixed with cold water to match particular preference.

Rainwater can be exclusively used for meeting the demand of hot water in a building. No extra measures are needed to heat and supply rainwater. In case of exclusive use of rainwater for hot-water supply, the requirement of water should be properly estimated. Table 4.9 can be used for estimating hot water consumption per occupant or person in common types of buildings.

Table 4.9 Hot-water consumption per occupant or person in common types of buildings [13]

Type of building	Consumption per occupant (litre/day)	Peak demand per occupant (litre/h)	Storage per occupant (litre)
Factories (no. of processes)	22–45	9	5
General hospitals	160	30	27
Mental hospitals	110	22	27
Hostels	90	45	30
Hotels	90–160	45	30
Houses and flats	90–160	45	30
Offices	22	9	5
Boarding schools	115	20	25
Day schools	15	9	5

Table 4.10 Water requirement for washing vehicles washing associated with a building

Sl no	Type of vehicles	Water requirement
1	Two-wheeled carriage	30–40 litres/day [3]
2	Four-wheeled carriage	50–70 litres/day [3]
3	Car (self-service operation)	55–82 litres/vehicle [14]

Vehicle washing: Building occupants might have vehicles for their use. These vehicles require water mostly for cleansing purposes. The amount of water needed for these purposes is quite considerable. Such use can be well met by rainwater. Estimating the water requirement for these purposes can be performed using the values listed in Table 4.10.

4.7 Loss and Wastage of Water

All water-supply systems and hydraulic equipment in buildings may experience some water loss predominantly due to leakage. Leaking of pipe joints may also cause water loss, which cannot be tolerated in building. Water loss is also called "unaccounted for water" to distinguish it from losses that occur for known reasons, such as the flowing of water that is not actually used. The amount of water unaccounted for in a water-distribution system is typically expressed as a percentage of the total amount of water pumped or drawn from the supply source. In estimating the water requirement for a building, this amount of water loss should be considered.

In a building, particularly in small low-rise residential buildings, there might be little water that is unaccounted for, but in large high-rise and residential buildings such losses can be <5 %. In office and public buildings where the pressure in the water-supply network is maintained high (>500 kPa), 10–30 % water loss due to leakage may be an unusual phenomenon [15].

4.8 Factors Affecting Per-Capita Demand

In tables of water requirements for various purposes, mostly a range of quantity is furnished. In estimating the quantity of water for a particular use, various factors affecting the demand should take a value either on the lower side or the greater side of the range or a value within the range. Following are the main factors affecting the per-capita demand for water in a building or large projects comprising various buildings.

4.8.1 Climatic Conditions

The quantity of water required in hot and dry areas is greater than that amount required in cold areas because of the use of various cooling amenities such as air coolers, air conditioners, etc., and other purposes such as sprinkling lawns, gardens, and courtyards, washing rooms, washing of clothes, and bathing etc. However, in very cold countries sometimes the quantity of water required may be more due to wastage, because of uncertainty, related to keeping taps open for fear of freezing the water in the pipes, thus causing continuous loss of water, and the use of hot water to keep rooms warm.

4.8.2 Size of Community

Water demand is greater with an increase in the size of buildings because more water is required for floor washing, car washing, running and cleaning of sewers, maintenance of open space and gardens, etc.

4.8.3 Living Standard of the People

The per-capita demand for water in a building increases with increasing standard of living of the dwellers because of the greater use of air conditioners, room coolers, lawn maintenance, use of automatic home appliances, and recreational purposes such as fountains, waterfalls, swimming pools, etc. Table 4.11 lists the variation in domestic-water consumption of different income groups of selected countries.

Table 4.11 Domestic per-capita consumption by different income groups of selected countries: Turkey, Saudi Arabia, Egypt, Indonesia, Hong Kong, and Bolivia

Housing class	Description	Water consumption litres/capita/day
High income	Detached houses, luxury apartments having ≥2–3 taps per household	150–260
Middle income	Houses and apartments having at least 1 WC and 2 taps per household	110–160
Lower income	Tenements, government housing, and shared houses having at least 1 tap per household but sharing WC	55–70

Source [16]

4.8.4 Manufacturing and Commercial Activities

A building, particularly high-rise or large building, may have mixed occupancy including various manufacturing and commercial units. Because the quantity of water required in those occupancies is much more than for domestic demand, their presence in a building will enormously increase the per-capita demand of the population of the building. In fact, the amount of water required by the particular type of manufacturing units must be determined by the activities where water is required and should be added in with the other water demands to estimate total water demand for the building.

4.8.5 Pressure in the Distribution System

The rate of water consumption automatically increases with an increase in pressure maintained in the water-supply system of the building. Generally in high-rise buildings, high pressure is maintained to supply the farthest point of use. Buildings having plumbing fixtures requiring comparatively high pressure to activate also need comparatively high pressure in the water-supply system. The consumption of water in a building incorporating a high-pressure water-supply system pressure will increase first due to the high rate of water flow compared with the flow in normal or low-pressure systems and more water loss due to leakage, wastage, and water flow that is unaccounted for, etc.

4.8.6 Sanitation and Drainage System

The water requirement in a building will vary depending on the building-drainage system adopted and the types of plumbing fixtures used. A building incorporating a "two-pipe" system will consume comparatively more water for cleaning and flushing building drains and sewers than a building incorporating a "one-pipe" system. A building using water-conserving plumbing fixtures—such as dual-flushing water closets, self-closing faucets, water-conserving shower heads, etc.—will consume comparatively less water.

4.8.7   Cost of Water

The cost of water directly affects its demand in a building. If the cost of water is high, less quantity of water will be consumed. In contrast, when the cost is low, water will be used lavishly, and more wastage will occur.

References

1. Butler D Memon FA (2006) Water demand management. LWA publication, London, p 4
2. Gleick PH (1996) Basic water requirements for human activities: meeting basic needs. Water Int 21:83–92
3. Jain VK (1985) Handbook of designing and installation of services in high rise building complexes. Jain Book Agency, India
4. Government of United Kingdom (2008) future water, the government's water strategy for England. State for Environment, Food and Rural Affairs, https://www.gov.uk/government/uploads/system/uploads/attachment_data/file/69346/pb13562-future-water-080204.pdf Retrieved on 14 Sep 2015
5. World Water Council (WWC) Water Crisis http://www.worldwatercouncil.org/library/archives/water-crisis/ Visited on 14 Sep 2015
6. Environment Agency UK (2008) Water and the environment. International Comparisons of Domestic Per Capita Consumption, webarchive.nationalarchives.gov.uk/20140328084622/http://cdn.environment-agency.gov.uk/geho0809bqtd-e-e.pdf Retrieved on 19 Jan 2016
7. USAID (U D) Office building water efficiency guide. Water Demand Management Program, p 9
8. Housing and Building Research Institute (HBRI) (1993) Bangladesh National Building Code 1993
9. Ministry of Community Safety and Correctional Services (1999) Fire protection water supply guideline for part 3 in the ontario building code. Office of the Fire Marshal, Ontario', http://www.mcscs.jus.gov.on.ca/english/FireMarshal/Legislation/TechnicalGuidelinesandReports/TG-1999-03.html#P101_13056 Retrieved on 04 Oct 2015
10. Birdie GS (1982) Water supply and sanitary engineering. Dhanpat Rai and Sons, Delhi Page 81
11. Yadav A, Patel P (2014) Assessment of water requirement and calculation of fire flow rates in water based fire fighting installation. Int J Innovations Eng Technol (IJIET), 4(1)
12. European Commission (2009) Study on water performance in buildings Bio intelligent service p 31
13. The Engineering Toolbox Hot Water Consumption per Occupant http://www.engineeringtoolbox.com/hot-water-consumption-person-d_91.html Retrieved on 23 Nov 2015
14. International Carwash Association (2002) Water use in the professional car wash industry. Report, p 34, www.carcarecentral.com

15. Commonwealth of Australia (2006) Water Efficiency Guide Office and Public Buildings. The Department of the Environment and Heritage, Canberra Page, p 11
16. Wagah GG, Onyango GM, Kibwage JK (2010) Accessibility of water services in Kisumu municipality, Kenya. J Geogr Reg Planning 3(5):114 Available online at http://www.academicjournals.org/JGRP Retrieved on 04 Oct 2015

Chapter 5
Rainwater Collection

Abstract To harvest rainwater in building, it is collected from the potential catchments of the building and its surrounding premises. The roof of a building serves as the most effective catchment, but in critical cases various other type of catchments are used. A collection system should be provided to collect rainwater from the catchments. Although roofs are mostly a flat and horizontal surface, roofs of various configurations and inclinations are also seen. As a result, the types of collection and collecting systems, due to roofs of varied shapes, configurations, and inclinations, also differ. The roofs of buildings are mostly made of reinforced cement concrete, but roofs of various other materials also exist. Roofing materials influence the quality of the collected rainwater. Various surface-finishing materials are sometimes used on roofs for various purposes, which also contribute to the quality of rainwater. After a considerable dry spell, the rainwater of the first rain is generally found to be comparatively more contaminated and polluted. Therefore, a considerable amount of such first rainwater must be flushed out. In this chapter, various catchments of buildings, particularly roofs of various designs and materials, as well as the collection approaches and first rainwater–flushing methodologies, are discussed. Determining size of the rainwater-collecting elements, the volume of rainwater to be collected, and the quality aspects of the rainwater, is also discussed. Finally, ways of maximizing rainwater collection are suggested.

5.1 Introduction

When it rains, raindrops fall there on all surfaces exposed to the sky. In the case of buildings, roofs are usually the largest surface exposed to the sky. Other than roofs, many elements on and outside of a building also remain exposed to the sky. After falling, some of the rain on those exposed built surfaces is absorbed by the surface materials; some rain may evaporate; and the rest of the rain falls outward or downward. Generally rainwater is collected from the roof of a building, but when rainwater from the roof is not sufficient, other catchments are considered for collection. The roof of a building is supposed to be flat for rainwater collection, but

© Springer International Publishing Switzerland 2017
S.A. Haq, PEng, *Harvesting Rainwater from Buildings*,
DOI 10.1007/978-3-319-46362-9_5

this is not mandatory for all cases. Various types of roofing elements are also designed in various styles and shapes, which influence the rainwater-collection system. In this chapter, the ways of collecting the rainwater falling on various roofs and other exposed surfaces of buildings is discussed.

5.2 Catchments

A catchment is a stretch of exposed surface area on which precipitation falls and flows downward toward the draining outlets. The volume and rate of rainwater runoff are the functions of the catchment area, the intensity and duration of rainfall, the slope of the surface, and the type of surface-coverage material. The collection system will depend on the location, size, shape, and accessibility of the catchments.

5.2.1 Building Elements as Catchments

In buildings, elements that remain exposed to open sky are the major concerns of rainwater harvesting because rainwater will fall on those exposed surfaces from where it can be collected for harvesting. Depending on the size and location of those elements and the rainfall intensity, collection pipes should be planned, sized, and installed. The building elements that remain fully or partially exposed to open sky are as follows.

1. Roof
2. Verandas and balconies
3. Sun shades and cornices
4. Car porch
5. Part of any side walls

Rain falling on these surfaces can be collected for harvesting. Therefore, the exposed surfaces of these elements can be designated as catchments for rainwater harvesting in buildings.

5.2.2 Planning the Catchments

Whenever any building element is chosen as a catchment for collecting rainwater, then it should be designed in such a way that helps in collecting rainwater. Particularly for horizontal surfaces, there should be parapet around the surface to prevent the freefall of rainwater. In unavoidable cases, a channel may be created before the edge line, as shown in Fig. 5.1, to receive and help the flowing toward

Fig. 5.1 Parapet and channel
on a roof to collect rainwater

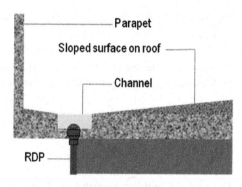

Fig. 5.2 Exposed horizontal
elements on wall

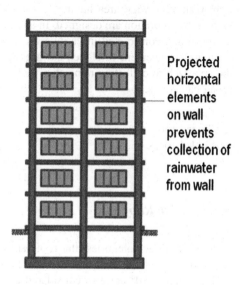

the inlet of collection pipes before the rainwater falls freely. In vertical surfaces to be used as catchment, there should not be any exposed horizontal elements as shown in Fig. 5.2, which can divert rainwater dripping down the wall surface [1].

5.3 Rainwater Collection from a Roof

In every building, the roof is the potential catchment from where a considerable amount of rainwater can be collected for its comparatively greater expanse of exposure to the sky. That is why the roof is the first choice as a catchment for rainwater harvesting in a building. The roof has many other advantageous features and characteristics to become the primary catchment for rainwater collection as follows:

1. Less possibility of becoming dirty due to its elevated position
2. Provision of easy accessibility
3. A wide spread, open, and free space
4. Maximum rainwater collection at minimum cost
5. Flexibility of storing rainwater at various levels in building.

5.4 Types of Roofs

The roofs of buildings are usually built as a flat surface. In special cases, particularly with very wide area having few numbers of or no support inside, roofs of various configuration are designed. In addition, for aesthetic purposes, top roofs are designed and constructed in various ways. The roof may be a single flat roof or a complex arrangement of sloped surfaces known as "pitched roof." Depending on the approach to collecting rainwater, roof types are grouped in the following three broad categories.

1. Flat roofs
2. Sloped roofs
3. Folded roofs.

5.4.1 Flat Roof

Among the three categories, flat roofs are unique and mostly designed to be flat. Flat roofs are usually constructed to be horizontal having a flat surface, but roofs with an angle <10° are also considered as flat roofs [2]. It is more economical to build a flat roof because it requires less construction, shuttering, and propping materials as well as shorter construction time. However, for drainage purposes, the top surface of flat roofs is usually slightly sloped. Therefore, sometimes roofs initially built flat are made sloped by providing a thin-layered, finished-sloped re-roofing surface.

Flat roofs with parapet all around are susceptible to the accumulation of rainwater on it, due to clogging of all the inlets of rainwater down pipes. Sudden collapse of flat roofs may happen due to having a pooled-water load exceeding the ultimate carrying capacity of the roof. Flat roofs are not an ideal choice for areas subject to heavy snowfalls.

5.4.2 Sloped Roofs

This group includes roofs having a sloped surface >10°. Rainwater falling on this type of roof rolls downward. Various roofs having sloped surfaces making different shapes are grouped under this category as follows.

1. Hipped roofs
2. Pyramid hip roofs
3. Pitch or gable roofs
4. Gazebo roofs
5. Mansard roofs
6. Shed roofs
7. Gambrel roofs

Hipped roof: A "hipped" roof or "hip" roof is a type of roof where all sides are sloped downward from the ridge to the outer supporting walls. The sloped sides of the roof are usually placed at a moderate slope featuring a ridge across the uppermost portion of the roof as shown in Fig. 5.3.

Pyramid hip roof: This is a type of hip roof for which there is no linear ridge. Here all four sloping roof surfaces meet at a single point. Figure 5.4 shows the pyramid hip roof.

Fig. 5.3 Hipped roof

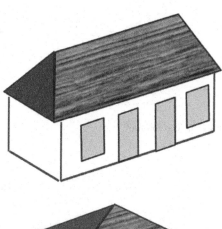

Fig. 5.4 Pyramid hip roof

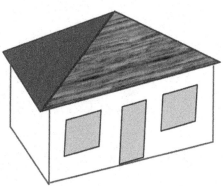

Pitch or gable roof: Gabled roofs have two sloping roofs joining together along a ridge creating a triangle at two ends. Therefore, the end walls are erected with a triangular extension, called a "gable", at the top of wall. Figure 5.5 represents a gable or pitch roof.

Gazebo roof: A gazebo is a combination of multiple triangular sections of roof, often octagonal roofed structure. There are also hexagonal gazebo roofs consisting of six triangular plates as shown in Fig. 5.6.

Mansard roof: A mansard roof, also called a "French" roof or a "curb" roof, is a four-sided gambrel-style hip roof having two slopes on each of its sides. In this type of roof, the lower part of each side has steeper angle than that of upper part as shown in Fig. 5.7.

Shed roof: A shed roof, also known as "lean-to" roof, is typically a single-plate roof that is placed inclined in one direction and slopes down the entirety of the structure or structure additions. It is generally the least expensive and easiest roof to build. No extra surface layer needs to be provided for drainage. This type of roof is also built on building additions, sheds, and porches. When such a roof is used as an addition to a building, the roof is generally kept supported on the building wall along the high side as a "lean-to." These shed roofs may also be attached to the

Fig. 5.5 Pitch or gable roof

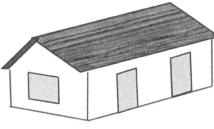

Fig. 5.6 Gazebo roof
(Hexagonal)

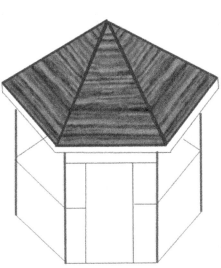

Fig. 5.7 Mansard roof

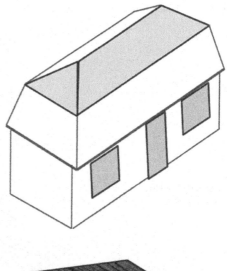

Fig. 5.8 Shed roof or lean-to roof

building supporting by walls, or it may stand alone supported by columns. Figure 5.8 shows a shed roof supported by walls.

Gambrel roof: A gambrel or gambrel roof is usually a symmetrical two-sided roof having two or three sloping of plates on each side. The slope of the upper plates is positioned at a shallow angle compared with the slope of the lower plate. The lowest plate of each side has the steepest slope as shown in Fig. 5.9.

5.4.3 Folded Roofs

Folded roofs are generally constructed by combining a series of sloped flat plates joined at various angles. Folded roofs create a gutter between two plates. There are various types of folded plate roofs as follows:

1. Sawtooth roof
2. M-shaped roof
3. Butterfly roof

Fig. 5.9 Gambrel roof

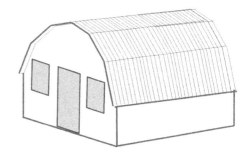

Fig. 5.10 Sawtooth roof

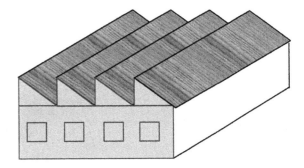

Saw-tooth roof: A roof comprising a series of ridges with dual pitches on either side that look like the teeth of a saw is termed a "sawtooth roof". This roof comprises of series of vertical glazed surfaces to allow light to enter into the building and opaque sloping surfaces that provide shade. In large single-storey factory buildings, saw-tooth roofs were popular for having letting more natural light inside. Figure 5.10 represents a saw-tooth roof.

M-shaped roof: This type of roof is similar to a double-pitched roof placed together creating a gutter in between as shown in Fig. 5.11.

Butterfly roof: A butterfly roof is shaped like the wings of a butterfly. A butterfly roof consists of two plates that dip down in the middle and slops upward at each end as shown in Fig. 5.12. The dip serves as a gutter.

5.5   Catchment-Surface Materials

The roof of a building is always considered as the first and most effective choice for the catchment of rainwater. But the fact is that the roof is the most vulnerable element of a building, amongst its other external elements, due to its direct exposure to weather and atmosphere. Therefore, various layers of different materials are put over the roof surface to protect it from weather events as well as to provide comfort in the indoor environment. Rainwater quality from different roof catchments is a

Fig. 5.11 M-shaped roof

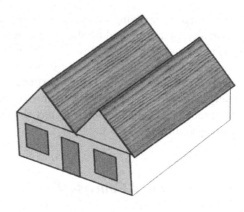

Fig. 5.12 Butterfly roof

function of the type of material on the roof surface, the climatic conditions, and the surrounding environment. Various types of materials used for roofing and finishing materials, as well as their effect on rainwater quality and quantity, are discussed below.

5.5.1 Metal Surface

The quantity of rainwater that can be collected from a roof is a function of the roof texture and the water-absorption capacity of the roofing material. A smoother surface and dense material result in more runoff and therefore a better volume of collection. Thin metal sheets, both smooth and corrugated, are usually used as roofing material. A commonly used roofing material that is also suitable for rainwater harvesting is a galvanized iron sheet, which is commonly known as a GI sheet. Iron sheets are also available with various colour coatings. The sheet can also be colored by applying an epoxy coating, which provides a hard chemical- and solvent-resistant coloured finish on the roof surface. Care must be taken in selecting paints for roofing surfaces from which rainwater should be collected. Lead-based paint must be avoided. Rainwater collected from roofs with copper flashings may cause discolouration of porcelain fixtures. One study reported that aluminum is suitable as a roofing material for rainwater harvesting because rainwater collected from roofs made of aluminum sheeting showed a high concentration of magnesium ranging from 36 to 38.6 mg/l [3].

5.5.2 Clay and Concrete Tile

Clay and concrete are easily available materials for making roofing tiles. These tiles are generally used in sloped roofs. The roof surface made of these clay or concrete tiles are suitable for both potable and nonpotable use of rainwater, but they may contribute to as much as a 10 % loss [4] due to their rough texture, friction provided during flow, porosity, and evaporation. Due to being porous and having a rough surface, dirt is accumulated in the corners of tile ribs, thus promoting the growth of various microorganisms. To decrease water loss and prevent the growth of microorganisms, tiles are painted or coated with a sealant. There is a chance that toxins may leach from the sealant or paint. This painted or coated roof surface is made safer by painting it with a special sealant or paint that contains few or no toxic ingredients in order to prevent bacterial growth on the porous surface material.

5.5.3 Composite or Asphalt Shingle

Asphalt-shingle roofing materials are made of tar-like hydrocarbon speckled with small, coloured ceramic granules. The mineral granules protect the roof from the ultraviolet (UV) light of sun, which deteriorates asphalt over time. Asphalt is made from a byproduct of refining crude oil, and the leaching of toxins from coloured composite shingles may not be appropriate for direct consumption of rainwater as potable water. Composite roofs generally result in decreased collection due to inefficient flow or evaporation loss, which is approximately 10 % of the amount received on the catchment [5, 6].

5.5.4 Other Roofing Materials

Lime terrace: Lime terrace is provided over a concrete flat roof to protect it from heat and to create a sloped surface for storm-water drainage. It is made of lime and brick aggregates. These materials have little or no adverse effect on rainwater quality.

Slate: Slate helps to create a very hard and smooth roof surface, which is ideal for a catchment because it maximizes the production and potable use of rainwater assuming that no toxic sealant is used. However, the cost of slate is high, thus limiting its use as a roofing material.

Plastic: Plastic sheets are also used as roofing material. Rainwater can be harvested from a roof made of plastic sheets because the quality of rainwater collected from it, particularly the pH and total hardness, falls within the acceptable limits of WHO recommendations [3].

Fiberglass: Laminated fiberglass shingles are a popular roofing material because they last longer than asphalt shingles. There is possibility for the fiberglass to leach into the rainwater, which makes this material a poor choice from the perspective of rainwater quality [7].

5.6   Rainwater Collection from a Roof

Roofs are constructed in a variety of shapes as described previously. The shapes vary from a single flat roof to multi-span pitched construction with a valley, ridge, parapet, or boundary wall etc., as shown in Fig. 5.13. The rainwater-collection system for roofs of buildings depends on the roofing system of the building. The collection of rainwater from a differently shaped roof is discussed below.

5.6.1   Collection from a Flat Roof

Rainwater falling on any flat roof surfaces is collected by providing a good number of properly sized rainwater down pipe (RDP) inlets at suitable locations, mostly around the periphery of roof. If the roof is not provided with any parapet wall to prevent the overflowing of rainwater, a curb should be made all around at the periphery of roof. To direct the flow of rainwater toward the inlets, a flat roof must be provided with another roofing layer, thus making its surface slope very gently

Fig. 5.13 Multi-span pitched construction with valley, ridge, parapet, and gutters

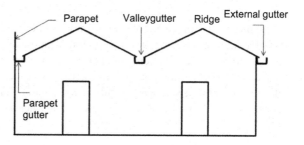

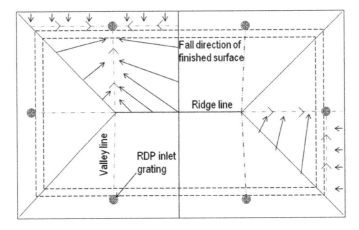

Fig. 5.14 Rainwater inlets and sloping of surface on a flat roof

toward the inlet. To direct the rainwater toward multiple inlets, multiple sloped surfaces are made in the same way. Each sloped surface for a particular rainwater inlet is configured depending on the location of that particular inlet, the number of inlets provided, and the shape of the roof. The configuration of sloped surfaces for rainwater down pipe inlets in a rectangular roof is shown in Fig. 5.14.

Sometimes it is required to connect two or more inlet pipes to one collection pipe leading to a rainwater-collection reservoir. This system requires a connection of inlets by pipes installed under the catchment, which may not be acceptable for various reasons. In this case, an extra false ceiling may be required to cover the exposed pipes under the roof or any other catchments as an aesthetic and protective measure.

Collection can be performed by creating a channel at the periphery of the catchment. The channel bed should be sloped toward a collecting point. At the collecting point, there should be two inlets, but the collection should be made by one pipe receiving rainwater from two inlets and leading to the reservoir for storage. During the design of roof, the depth of channel should be estimated properly.

5.6.2 Collection from a Sloped or Curved Roof

The approach of rainwater collection from a sloping-surfaced roof is different from that of a flat roof. For sloped roof surfaces, a gutter is fabricated under the sloped or curved surfaces. A sufficient number of rainwater inlets are placed at suitable locations. The flow direction of rainwater in the gutter can be pointed toward any inlet by installing a segment of gutters for a particular inlet and installing it at a proper slope with rainwater falling toward the inlet as shown in Fig. 5.15.

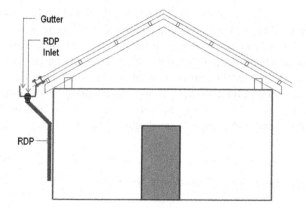

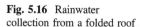

Fig. 5.15 Rainwater-collecting gutter under a sloped roof

Fig. 5.16 Rainwater
collection from a folded roof

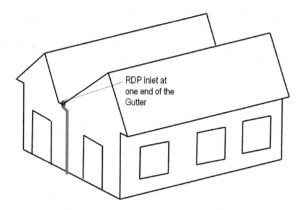

5.6.3   Collection from a Folded Roof

In folded or pitched roofs, valleys are created in between two plates toward which the rainwater flows downward. A rainwater inlet is placed at the end of valley if the contributing catchment is sufficient to be served by one inlet of considerable size as shown in Fig. 5.16. When the length of the valley runs too long, and the contributing catchments cannot be served by one inlet of considerable size, then two inlets should to be provided at both ends of the valley, thus making a ridge at the center.

5.7   Effective Catchment Area

The effective catchment area for roofs falling under various groups will not be the same even though the plan of the surface area of the roofs is the same. An effective catchment area depends on the effective area of the roof that is vertically exposed to the sky. Therefore, the effective catchment area predominantly depends on the inclination of the roof. The greater the inclination of the roof, the greater the reduction in effective catchment area even though the total surface area of the roof remains the same. Therefore, in determining the volume of rainwater to be collected or drained, the effective catchments of the roof and the wind-driven rain running off the adjacent roofs and adjoining wall, if any, contributing to the accumulation of rainwater on the roof surface must be considered. Determination of the effective catchments of flat and inclined roofs are discussed below.

5.7.1 Catchments of a Flat-Roof Surface

For a flat-roof surface, the effective catchment area is its horizontal plan area plus 50 % of the one adjoining vertical wall, which contributes to rainwater accumulation on the concerned catchment.

Let us consider a building having a flat roof at different levels as shown in Fig. 5.17. In level 1, the plan area ABCD is the catchment contributing to the rainwater down pipe RDP1.

Fig. 5.17 Catchments of flat-roof surfaces [8]

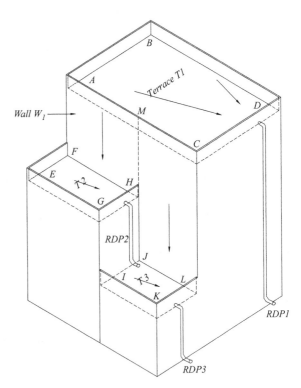

In level 2, the horizontal plan area EFGH and 50 % of the adjacent vertical wall surface area AFHM projecting above are the catchments contribute rainwater flowing toward rainwater down pipe RDP2.

For the rainwater down pipe RDP3, the catchments will be the plan area of terraces at level 2 and 3 plus 50 % of the area of the adjacent contributory vertical walls AFHM and MJLC.

5.7.2 Catchment of an Inclined Surface

In determining the total effective catchment area of the inclined roof on a building shown in Fig. 5.18, the following areas should be calculated.

1. The horizontal plan area
2. Fifty per cent of the vertical elevation area
3. Fifty per cent of the adjacent wall area

Let us consider an inclined roof ABCD. Its horizontal plan area is ABC'D'; the vertical elevation area is CDC'D'; and the adjoining wall ADE contributes to the accumulation of rainwater on this roof.

The total effective catchments for the roof of the building will thus be the plan area ABC'D' plus half of the elevation area CDC'D' plus half of the wall ADE.

Example:
Let us consider an inclined roof ABCD of Fig. 5.18 having a length of 25 m, width of 10 m, and pitch height of 5 m situated just adjacent to a side wall, which will also contribute in generating rainwater for collection. Then the total effective catchment area will be as follows.

$$\text{Area} = \text{roof plan ABC'D'} + \frac{1}{2} \text{ beveled surface CDC'D'} + \frac{1}{2} \text{ side wall ADE}$$
$$A = (25 \times 10) + (5 \times 25)/2 + (10 \times 5)/2$$
$$= 250 + 62.50 + 12.50$$
$$= 325 \text{ m}^2$$

Fig. 5.18 Catchments of sloped-roof surfaces

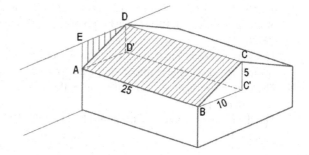

5.8 Rainwater-Conveying Media

Rainwater flowing down on the catchments should be directed toward the rainwater-conveying elements to transport it to the desired destination. The effective media used in conveying rainwater in buildings are as follows:

1. Gutter
2. Rainwater down pipe (RDP).

5.8.1 Gutter

A gutter is a narrow channel that collects rainwater generally from any inclined roof of a building or any inclined surface and diverts it to the inlet of a collection pipe.

Gutter types: Depending on the position of the gutters and the surrounding conditions, gutters are classified as follows.

3. External gutter
4. Parapet or boundary wall gutter
5. Valley gutter

A ridge is a horizontal line formed along the elevated end by the juncture of two sloping planes. In a pitch roof, the ridge line is formed by the sloping surfaces at the top of the roof. From the ridge at top of the inclined catchment, rainwater flows downward to the gutters below. The inclination of the gutter directs the rainwater to the inlets of the collection pipes.

Gutter design: To determine the shape and size of a gutter correctly, it is necessary to calculate the rainwater discharge rate from the roof. This needs to assess the rainfall rate and the effective catchment area from which the rainwater should be collected.

The gutter has to be designed to provide sufficient capacity for the predicted discharge rate. Following assumptions are generally considered in the design of gutters.

1. The rule of thumb is that a gutter of 1-cm^2 cross-section is required for a 1-m^2 roof surface [9]
2. Gutters shall fall toward the collection pipe. The slope of the gutter should be <1 in 350 [10]. A steeper slope makes for a greater flow rate. By steeping the slope of the gutter from 1:100 to 3:100, potential rainwater flow can be increased by 10–20 % [9]
3. The gutter should have a uniform cross-section and should be large enough to ensure free discharge without spillage of rainwater.
4. The dimension from a stop end to an outlet should be <50 times the maximum water depth [10].
5. The distance between outlets should be less than 100 times the maximum water depth [10].

Table 5.1 Size of semi-circular gutter[a]

Diameter of gutter (mm)	Maximum projected roof area for semi-circular gutter of various slopes			
	5.2 mm/m	10.4 mm/m	20.8 mm/m	41.7 mm/m
76	15.8	22.3	31.6	44.6
102	33.4	47.4	66.9	94.8
127	58.1	81.8	116.1	164.4
152	89.2	126.3	178.4	257.3
178	128.2	181.2	256.4	362.3
203	184.9	260.1	369.7	520.2
254	334.4	473.8	668.9	929.0

Source [11]

[a]Data in Table 5.1 are based on a maximum rainfall of 102 mm/h for 1-hour duration. The value for the drainage area should be subject to rainfall in millimeters per hour of local conditions

The depth of flowing water in the gutter is not constant but rather varies: The depth is maximum at the upstream end and the minimum depth, also called the "critical depth," is at the outlet. The depth of water flow is dependent on the shape and slope of the gutter. For a rectangular-shaped gutter section, the maximum depth of flowing water is equal to the twice the depth at the outlet [10]. In addition to the depth of water flow, all the gutters should include an allowance in depth as a freeboard to prevent splashing and to allow rainwater flowing below the spillover level of the gutter. A minimum 50-mm freeboard is often considered a good practice [10].

The size of a semi-circular gutter should be based on the maximum projected roof area according to the Table 5.1.

5.8.2 Rainwater Down Pipe (RDP)

Rainwater down pipes are pipes that direct rainwater down from the roof or any other elevated surface to a lower-level surface or to the ground. A rainwater down pipe is also termed as "leader." The rainwater downpipe has an inlet at the top, which is kept covered by a grating, and the other end is kept open. The inlet of a vertical leader on any open surface should be provided with a dome-shaped grating as shown in Fig. 5.19. The size of the inlet of any vertical RDP should be one size larger than the size of the corresponding vertical RDP.

For collecting rainwater from any elevated catchment, the RDP can be used as an inlet pipe to the rainwater storage tank or any reservoir.

Sizing and determining the number of RDPs needed: The size and number of vertical leaders or rainwater down pipes should be based on the maximum projected

Fig. 5.19 Dome-shaped grating on the inlet of a rainwater down pipe

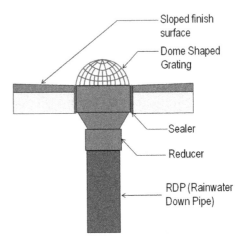

Sloped finish surface

Dome Shaped Grating

Sealer

Reducer

RDP (Rainwater Down Pipe)

roof area according to Table 5.2. A minimum of two roof drains should be provided for any independent roof surface. The minimum diameter of an RDP shall not be <50 mm. In a building, a rainwater down pipe can rarely be installed vertically throughout its length. In the majority of cases, the inlet is kept vertical, and the down pipe is offset either at the top or at the bottom. Sometimes, at an intermediate level, the rainwater down pipe needs to be offset. In these conditions, the capacity of rainwater down pipe is decreased. Therefore, it is wise to consider the capacity of the rainwater down pipe having both horizontal and vertical alignment as a that of a horizontal rainwater-conveying pipe. The diameter of a horizontal rainwater-conveying pipe in millimetres should be based on the maximum pro-jected roof area (m^2) and the intensity of rainfall (mm/h) according to Table 5.2. The slope of horizontal portion of pipe is considered to be 100:1 [12].

Locating RDP: After determining the number of rainwater down pipes needed, their positioning should be well planned in accordance with the configuration of the roof plan and the position of the other catchment areas of the building. Where possible, rainwater down pipes should be proportionately distributed along all sides of a building. Rainwater down pipes should not be installed too far away from the ridge such that at ridge the thickness of the finishing layer on the roof become too thick.

5.9 First-Flush Diversion

The roof and other catchments of building are subject to the deposition of dusts, leaves, bird droppings, and other debris. When it rains after a long dry period, those accumulated items are washed off the roof surface, thus making the rainwater dirty and contaminated. After a certain quantity of rain has fallen on a surface, the loose dirt and debris is more or less washed away, and the rainwater flowing afterward is

Table 5.2 Size of vertical leaders[a]

Size of RDP[b] (mm)	Maximum Projected Roof area and Flow	
	Roof area (m^2)	Rainwater flow (lpm)
50	202	87
65	367	155
75	598	253
100	1287	544
125	2336	986
150	3790	1602
200	8180	3450

Source [12]

[a]Data in Table 5.2 are based on a maximum rainfall of 25 mm/h for a 1-hour duration. The figure for the drainage area should be adjusted to local conditions of rainfall

[b]The equivalent diameter of square RDP will be the diameter of the circle that can be inscribed within the cross-sectional area. The equivalent diameter of the rectangular RDP will be the short dimension of the rectangular RDP. However, the ratio of width to depth of the rectangular RDP should not exceed 3:1

assumed to be of far better quality compared with the rainwater generated initially. Therefore, in rainwater collection an arrangement should be made in the rainwater leader or down pipe to divert the first washings from the roof or terrace catchments. The quality of the harvested rainwater from the catchments generally improves with flushing of the first rain for some duration as the rain event progresses. Treatment or conditioning of rainwater emphasizes the importance of an effective first-flush diverter in decreasing concentration of contaminants in the rainwater before collection. As a rule of thumb, it is believed that the contamination in rainwater runoff is halved by each millimeter of first-flush rainfall discarded [13]. Because first-flush diversion is a simple way to improve the quality of rainwater, like all other technologies it must be properly managed and maintained.

5.9.1 Sizing the First Flush

The duration of first flushing depends mainly on the air pollution in the area of concern. Again the larger the roof and the longer the periods between rainfall events, the larger the quantity of rainwater that must be disposed of through first-flush. There are a number of guide lines in these contexts as follows:

1. After 3 consecutive days rainfall of <1 mm, the first runoff of 1-mm depth should be diverted before collection [14]. According to Thomas and Martinson, a maximum 8.5 mm of the first flush has been recommended for targeted

turbidity of 5.0 NTU from 2000 NTU and minimum of 1 mm of the first flush for targeted 50 NTU from 100 NTU of initial turbidity of rainwater [15]

2. According to the Water Services Association of Australia (2005), it is recommended that the first-flush divert should be a minimum of 10 litres/100 m^2 of water before rain enters the rainwater storage tank [16] where the collection area is the area of the roof footprint

3. The first flush–diversion time should be within the range of a minimum of 5 min [17] to a maximum of 20 min [18].

5.9.2 Methods of First-Flush Diversion

The systems adopted in diverting the first flush of rainwater are broadly grouped into two systems: manual and automatic. There are again varieties of both automatic and manual diversion systems. Here only the basic principles of both the systems are discussed. For long-term use, a simple system created from locally available parts is the best option.

5.9.3 Manual First-Flush System

The simpler of first-flush diversion systems is based on a manually operated arrangement where a diversion pipe is provided on the rainwater-collection pipe leading to the rainwater storage tank as shown in the Fig. 5.20. On both the diversion and collection pipe, a gate valve is installed. When the rain starts, the gate valve of the collection pipe is closed while the gate valve on the diversion pipe is open. When the initial first flush has been diverted for suggested period, the gate

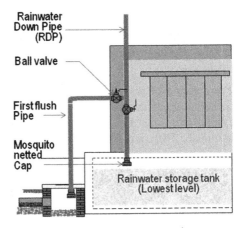

Fig. 5.20 The manual first-flush system

valve is closed, and the gate valve on collection pipe is opened to collect subsequent rainwater. This method has a major drawback because a person must be present who will perform these jobs at the start of the rain as well as after some time after it has started raining.

5.9.4 Floating-Ball First-Flush System

The floating-ball first-flush system is an automatic system. In this system, the first rain to flow through the collection pipe is accumulated in a collection chamber having a conical top as shown in Fig. 5.21. As the chamber fills with rainwater, a ball floats on the collected rainwater. As the level of collected rainwater rises, the ball also rises. Eventually the ball becomes stuck in the entrance passage between the conical top and the collection chamber thereby closing the collection chamber. When the passage of the collection chamber is closed by the ball, subsequent incoming rainwater is redirected to the diversion outlet leading to the main rainwater storage tank. The volume of the collection chamber should be large enough to store the first rainwater to be flushed out.

The first-flush diverter's chamber size is based on the desired amount of runoff to be diverted. If the diversion height is typically considered to be 1 mm, then the following is true:

First-flush diverter's volume (litre) = Diversion height 1 mm × Catchment area in m².

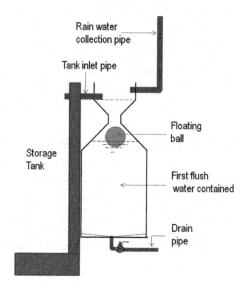

Fig. 5.21 The floating-ball first-flush system

5.10   Techniques for Maximizing Rainwater Collection

Rainwater harvesting is done when the availability of water from conventional sources becomes limited. When dependency on rainwater increases, it becomes evident that the collection of rainwater must be maximized. For this purpose, it is necessary to apply various techniques as mentioned:

1. Increasing the catchment area
2. Choosing the finished surface
3. Planning the collection elements
4. Designing the collection elements
5. Choosing appropriate collection elements
6. Good workmanship.

5.10.1   Planning the Collection Elements

While making plan of a building, all sorts of catchments should be well addressed in planning of a rainwater-harvesting system in the building. Therefore, to maximize rainwater collection care should be taken so that the effectiveness of the catchments is not hampered due to poor planning of exposed building elements related to rainwater collection. The effectiveness of catchments depends on the exposed free space around and above any catchments. Therefore, in some cases an increase of catchments might not be effective in maximizing the rainwater collection due to the hindrance of various elements around and above the catchments.

The length of the collection pipe from inlets to the collecting reservoir should be kept as short as possible. Greater length of collection pipe will require more joints where there is a possibility of an increased number of leaking joints, thus minimizing the volume of collected rainwater. Increased pipe length will cause increased friction loss resulting in reduced flow, i.e., reduced rainwater collection.

5.10.2   Increasing the Catchment Area

Flat catchments, from which rainwater is to be collected, can be increased in area as shown in Fig. 5.22. Therefore, the volume of rainwater falling on the increased surface will result in an increased volume of rainwater that could be collected. Similarly, on very sloped or angled surface, if the inclination is increased keeping the plan area of inclination same, the volume of rainwater falling on the increased

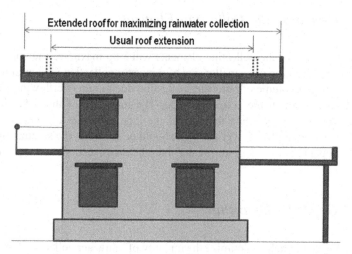

Fig. 5.22 Over-extended roof, compared with a usual roof, for increasing the catchment area

inclined surface will increase, which will maximize the volume of rainwater collection. Accordingly if the ridges of folded slabs are heightened, the area of inclined surfaces will increase, which will increase the volume of rainwater to be collected.

5.10.3 Choosing the Finished Surface

The finished surface of the catchments plays an important role in the generation of rainwater that can be collected. If the finished surface has high water-absorbing capacity, then more rainwater will be absorbed by the surface material and thus lost from collection. In this perspective, a metallic surface is better than a concrete surface. A concrete surface will be again better than any clay-tiled surface. The water-absorbing phenomenon of a concrete surface can be eliminated by laying ceramic tiles on it, which have a very poor water-absorbing capacity.

5.10.4 Choosing the Appropriate Collection Elements

Among the rainwater-collecting elements, the first-flushing device plays an important role in reducing the loss of rainwater collection. In a manual system, the operator may not operate the diversion outlet in a timely fashion, which will result in loss of collecting rainwater. In an automatic first-flush system, if the size of container for first-flushed water accumulation becomes larger, there will be a loss in the volume of collected rainwater.

5.10.5   Designing the Collection Elements

Appropriate design of the collecting elements will help maximize the collection of rainwater. Incorporation of a smaller size of rainwater-collection pipe or gutters will result in a reduced volume of rainwater collection due to rain overflowing from and decreased flow through the collection pipe. The inlet size of the collection pipe should be one size larger than the size of the pipe. The size of openings of the dome-shaped grating placed over the inlet should be such that the summation of the area of the openings is 2–3 times greater than the cross-sectional area of the inlet.

5.10.6   Good Workmanship

Good workmanship helps maximize the quantity of rainwater collected. Particularly good workmanship should be ensured in jointing pipes and with various others appurtenances and equipments for preventing leakages. There is every chance of wastage of rainwater through leaking pipe joints, cracks or holes in gutters, etc., thereby causing a decrease of rainwater collection.

References

1. Haq SA (2014) Rainwater harvesting in building: techniques of maximizing collection. Presentation in the 1st Regional Seminar on Climate change, water security and prospects of rainwater in Bangladesh, Organized by Military Institute of Science and Technology and Water Aid Bangladesh in August, 2014
2. Roofs (2015) Types of roofs. http://bbstore.northbrook-online.ac.uk/store/CITB/CnJ/materials/CJ12_Roofs/M121.pdf. Retrieved on 21 Sept 2015
3. Olaoye RA, Olaniyan OS (2012) Quality of rainwater from different roof material. Int J Eng Technol 2(8):1417. http://iet-journals.org/archive/2012/august_vol_2_no_8/287181339673653.pdf. Retrieved on 21 Sept 2015
4. Quality-drinking-water.com (2010) Rainwater collection system to harvest the rainwater. http://www.quality-drinking-water.com/rainwater_collection.html. Retrieved on 22 Sept 2015
5. Radlet J, Radlet P (2004) Rainwater harvesting design and installation workshop. Save the Rain, Boerne (TX)
6. Texas Water Development Board (TWDB) (2005) The Texas manual on rainwater harvesting, 3rd edn. Austin, Texas
7. Nate Downey (2010) Harvest the rain. Sunstone Press, USA, pp 11–112
8. Haq SA (2006) Plumbing practices. Syeda Masuda Khatoon, Uttara, Dhaka
9. Worm J, van Hattum T (2006) Rainwater harvesting for domestic use. Agromisa Foundation and CTA, Wageningen, The Netherlands
10. Kingspan (2015) Sandwitch panel India, Building design. http://panels.kingspan.in/Roof-Drainage-%7C-Roof-Drains-%7C-Gutter-Layout-%7C-India–13266.html. Retrieved on 22 Sept 2015
11. IAPMO Plumbing Code and Standards, India (2007) Uniform plumbing code—India 2008. IAPMO Plumbing Code and Standards Pvt Ltd., India

12. Housing and Building Research Institute (HBRI) (2015) Final draft, Bangladesh National Building Code 2015. https://law.resource.org/pub/bd/bnbc.2012/gov.bd.bnbc.2012.08.07.pdf. Retrieved on 21 Jan 2016
13. Martinson DB, Thomas T (2005) Quantifying the first flush phenomenon. In 12th International Rainwater Catchment Systems Conference, Nov 2005, New Delhi, India, Cited in Kelly CD (2008) Sizing the Firs Flush and its effect on the storage—reliability—yield behavior of rainwater harvesting in Rawanda' MSc thesis at MIT
14. Kelly CD (2008) Sizing the first flush and its effect on the storage-reliability-yield behavior of rainwater harvesting in Rwanda. Thesis for Master of Science, Massachusetts Institute of Technology, http://dspace.mit.edu/handle/1721.1/44289#files-area. Retrieved on 20 Jan 2016
15. Thomas TH, Martinson DB (2007) Roofwater harvesting: a handbook for practitioners. IRC International Water and Sanitation Centre, The Netherlands. Cited in Kelly CD (2008) Sizing the Firs Flush and its effect on the storage—reliability—yield behavior of rainwater harvesting in Rawanda' MSc thesis at MIT
16. Department of Planning and Local Government (2010) Water sensitive urban design technical manual for the greater Adelaide Region. Government of South Australia, Adelaide Government of South Australia, pp 5–23. https://www.sa.gov.au/__data/assets/pdf_file/0012/14142/WSUD_chapter_5.pdf. Retrieved on 20 Jan 2016
17. Worm J, van Hattum T (2006) Rainwater harvesting for domestic use. Agromisa Foundation and CTA, Wageningen, The Netherlands, p 62 (P)
18. Center for Science and Environment (CSE) (2003) A water harvesting manual' for Urban Areas' Case Studies from Delhi, New Delhi, p 10

Chapter 6
Rainwater Storage

Abstract Rainfall is not a continuous phenomenon. Therefore, to make rainwater available when there is no rain and there is dearth of water, rainwater must be collected and stored. A rainwater storage tank is one of the vital elements of rainwater harvesting in a building. Both the quantitative and qualitative aspects of rainwater harvesting are primarily dependent on the storage of rainwater. In developing appropriate and effective storage of rainwater, various factors, such as location, planning, size, aesthetics, and functionality of storage tanks, should be considered. The location of a rainwater-storage tank in a building must be judiciously decided due to having both the advantages and disadvantages of one particular choice out of multiple choices of locating the tank within or outside the building. It is important to plan how the rainwater will be stored, in conjunction with the water collected from other sources subject to the purpose of use considering the quality and quantity of water from different sources. The size of the storage tank is another factor to be decided giving the importance of the demand for rainwater, availability of rainwater, space for storage, and, above all cost of the tank. When tanks should be placed exposed, aesthetics may become a governing factor in deciding the shape and material of the tank. The functionality of the tank can be addressed by the reliability of the quality of stored rainwater. In this chapter, all aspects of rainwater storage are discussed focusing predominantly on the various options for locating the storage tank and identifying the advantages and disadvantages of all of the options.

6.1 Introduction

It is well understood that the process of raining in any area is not a continuous phenomenon. Its spatial occurrence is variable in nature depending on the place's location on globe. The use of water is frequent and continuous throughout life.

© Springer International Publishing Switzerland 2017
S.A. Haq, PEng, *Harvesting Rainwater from Buildings*,
DOI 10.1007/978-3-319-46362-9_6

Therefore, rainwater must be stored for the period when there will be no or little rain. The criteria for storing rainwater again depend on the water demand and rainfall characteristics. In this chapter, all of the concerns of rainwater storage inside or outside of a building are delineated.

6.2 Storing Rainwater

Storage of rainwater must hold rainwater that is collected during a rainy or wet period to use during dry periods when there is a scarcity of water from the usual available sources. In a building, rainwater is stored in reservoir or tank. A rainwater-storage tank is one of the major components in rainwater harvesting in buildings. Both the quantitative and qualitative aspects of rainwater harvesting are primarily dependent on the storage of rainwater. Appropriate and effective storage is the key for the success of rainwater harvesting. Various multidimensional factors should be considered for the development of a proper and cost-effective storage component of the harvesting system.

6.2.1 Factors of Storage Development

In developing appropriate and effective storage of rainwater, the following factors should be considered for successful implementation of a rainwater-harvesting system in any building.

1. Location
2. Planning
3. Size
4. Aesthetics and
5. Functionality.

6.3 Location of Storage

Rainwater can be stored at various locations of a building depending on technological suitability, safety, and economy. It can be stored inside as well as outside of the building. Storing rainwater in both locations has many advantages and disadvantages. It is generally preferred to locate the storage reservoir inside the building. In unavoidable circumstances, it can be located outside of the building but should be as close to the building as possible. Various factors should be considered in locating a storage tank as follows:

1. The space requirement for the size of tank needed
2. Structural safety of building elements supporting the tank
3. Availability of sufficient room to accommodate the height of the tank
4. Height restriction of the building
5. Accessibility
6. Limitation of pressure in the supply or distribution system
7. Exposure to sunlight
8. Atmospheric temperature: Heating and freezing of rainwater
9. Site constraints including conflicts with other utilities, future expansion, etc.
10. Aesthetics: Hide or highlight the tank.

These factors should be well addressed and considered while planning the location of the reservoir for rainwater storage with a view toward developing a sustainable rainwater-harvesting system in a building.

6.3.1 Storage Inside the Building

Inside the building, rainwater can be stored in various locations. Storing rainwater in all of these locations also has both advantages and disadvantages. Planning for defining the location of a storage tank should be performed during the architectural-planning stage of the building. The usual location of the reservoirs are either on the top or the bottom level of the building; however, storage can be located at any intermediate level, but sometimes it may become mandatory for technical reasons, particularly for large high-rise buildings.

Storing rainwater on top of a building: If the roof of a building is flat, then a reservoir can be built under the roof as shown in the Fig. 6.1. Even the roof on a

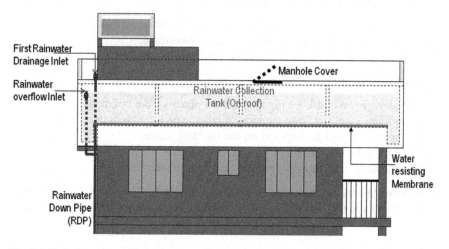

Fig. 6.1 Rainwater stored just under the roof on top of the building

stair or lift well can be used in the same way for a comparatively small amount of rainwater storage. Generally in low-rise buildings, where earthquake consideration is not very significant, rooftop storage can be easily performed.

The advantages of storing rainwater on top of a building include the following:

1. The length of collection inlet pipe will be shorter.
2. In some cases, rainwater can be collected directly without a collection pipe.
3. Arrangement for lifting water to the rooftop tank will not be required,

The disadvantages of storing rainwater on top of building include the following:

1. An extra load will be imposed on the building frame.
2. The building code may not allow it in particular areas given earthquake severity.
3. Extra precaution is required in structural design to manage slosh-dynamic problems related to large volume of stored water in an elevated tank.
4. Where there is restriction in the height of the building, extra height for a storage reservoir may not be allowable.
5. Supplying rainwater under gravity may not satisfy the pressure requirement in faucets of the upper-level floors.
6. There is chance of receiving sunlight inside the reservoir.
7. Automatic first-flushing may not be accommodated and cost-effective.

Storing rainwater at intermediate levels of a building: In some cases, rainwater can be stored at intermediate levels of a building. In unavoidable circumstances, particularly in large high-rise buildings, when the static water pressure in the piping at any intermediate level exceeds the maximum allowable limit, rainwater should be stored at a suitable intermediate level. The maximum static water pressure in the pipe should not exceed 552 kPa [1]. Storing rainwater at an intermediate level of a building, as shown in Fig. 6.2, has many advantages and disadvantages also.

The advantages include the following:

1. Rainwater can be stored from the roof and top catchments under gravity.
2. Water quality can be easily and frequently monitored.
3. Pressure in the supply system can be easily maintained.

The disadvantages include the following:

1. The storage tank will occupy usable floor space of the building,
2. Sound and vibration from pumping might create a problem in adjacent floors.
3. An extra load will be imposed on the building frame.
4. There may be chance of receiving sunlight inside the reservoir.
5. Maintenance activities may hinder other normal activities.

Storing rainwater at the lowest level of a building: Generally it is preferred to store rainwater at the lowest level of the building for various advantageous reasons.

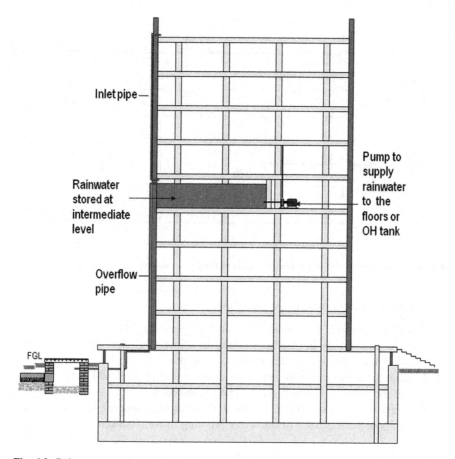

Fig. 6.2 Rainwater stored at an intermediate level of the building

However, there are also some disadvantages of storing rainwater at the lowest level of building as shown in Fig. 6.3. The advantages are as follows.

1. Rainwater can be stored from the roof and various other catchments under gravity.
2. No extra load will be imposed on the building frame.
3. The storage structure can be constructed at a comparatively lesser cost.
4. There is little or no chance of receiving sunlight inside the reservoir.
5. Construction of the reservoir's structural elements as an integral part of the building's foundation may improve the structural integrity of the building's substructure.

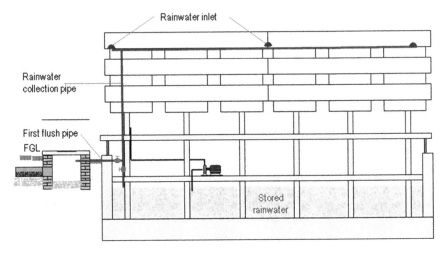

Fig. 6.3 Rainwater stored in the lowest basement floor of a building

The disadvantages include the following:

1. Extra pumping will be needed to supply the rainwater.
2. Extra pumping may be needed to drain overflow rainwater when the storage tank is built below the surface drains or sewers.
3. Leakage from the reservoir can cause deterioration of the load-bearing properties of the soil that supports the building foundation
4. Capacity of the load-bearing structural elements of a building may deteriorate due to poor performance of the water proofing measures taken.

6.3.2 Storage Outside of the Building

For rainwater harvesting in a building, the storage of rainwater can be located outside of the building when there is limited space inside but sufficient space outside for that purpose. The storage reservoir again can be built (1) above ground or underground or (2) partially above ground and underground as shown in Fig. 6.4. Storing rainwater outside of the building has both advantages and disadvantages.

Advantages of storing rainwater outside of the building include the following:

1. Building floor space that would be occupied by the reservoir could be used for other purposes.
2. There is no extra load on the frames of the building.
3. Leakage or overflow from tank will not make the floor wet or flood any part of the building.

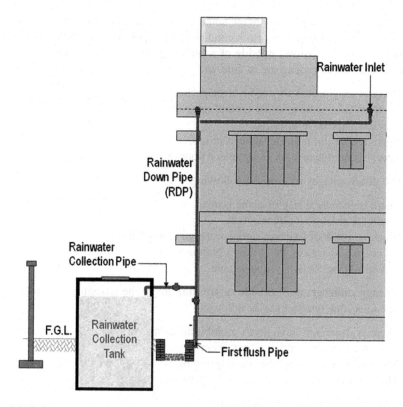

Fig. 6.4 Rainwater stored in a tank outside of a building

4. Installation of pump may also be outside the building, which will prevent problems of sound and vibration for the occupants.
5. There is increased flexibility in choosing the height and area of the reservoir, which may be limited in case of placing the reservoir inside the building.
6. The reservoir location can be constructed at a suitable time during development of plumbing system, thus allowing flexibility to change the reservoir size and construction material.

Disadvantages of storing outside the building include the following:

1. There may be chance of sunlight reaching the tank.
2. Green space will be covered, thus preventing the percolation of rain.
3. Depth may be limited for problems related to deep excavation.
4. Another structure for a pump house may be required outside.
5. There are risks of weather events.

Installation of a rainwater-storage tank outside of the building can be performed either under or on the ground. Both conditions have advantages and disadvantages.

Storage on the ground outside of a building: The advantages of locating the storage tank above ground are as follows.

1. There is easy access for repair or inspection.
2. There is little or no cost for ground excavation.
3. It is less expensive to install.
4. There are no groundwater-related issues.
5. There is little chance of going under flood water.

The disadvantages include the following:

1. There is increased risk of algae growth.
2. There are risks of weather events.
3. There is a risk of freezing stored water in cold climates.
4. It occupies space on the building premises.
5. It might be aesthetically undesirable.

Storage underground outside of the building: The advantages of locating a storage tank underground are as follows:

1. There is decreased daylight to prohibit algal growth.
2. It is protected from weather conditions.
3. It saves clear space on the building premises.
4. There is easy access for repair or inspection.

Disadvantages include the following:

1. The installation cost is relatively greater due to excavation.
2. It is less accessible for maintenance and inspection.
3. It requires a suitable location.
4. There is a risk of uplifting of the tank when it is emptied.
5. The depth may be limited due to various underground structural elements.

6.4 Planning of Rainwater Storage

To supply rainwater in a building, the storage of rainwater—with respect to its quality compared with that of the main water and the purpose of use—must be planned following one of the options given in Table 6.1. Here main water is considered to be the supplied water by the water-supplying authorities or any other establishments applicable.

Table 6.1 Planning options for the storage of rainwater with respect to its quality and purpose of use

Main water quality	Purpose of use	Planning of storage and conditioning
Directly drinkable	Drinking, cooking, bathing, ablution, etc.	Option 1: Main water and rainwater are stored separately (separate supply system) Option 2: Main water (drinkable) and rainwater (filtered and disinfected) are stored together in the same reservoir
Not directly drinkable	Cooking, bathing, clothes washing, etc.	Main water and filtered rainwater are stored together in a reservoir. Disinfect stored water separately for use
	Flushing, gardening, etc.	Main water and rainwater are stored together in a reservoir

6.4.1   Separate Storage and Supply System

Separate storage of rainwater and main water may have one storage-tank unit with a separating wall as shown in Fig. 6.5. Separate storage of rainwater and main water needs separate pumping and separate piping. On top of a roof, the tank may be one unit with a separating wall for keeping rainwater and main water separate as shown

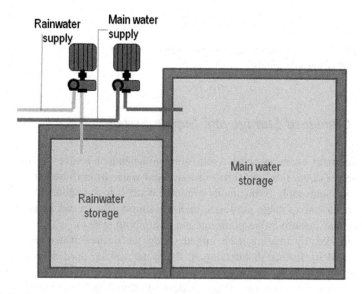

Fig. 6.5 Separate storage tanks for storing rainwater and main water separately

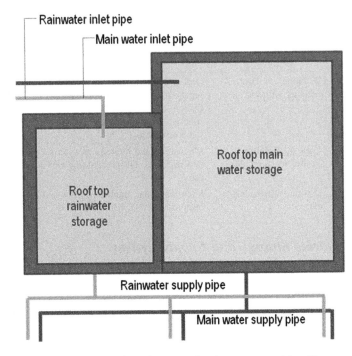

Fig. 6.6 One unit rooftop tank storing rainwater and main water separately with a separate piping system

in Fig. 6.6. Otherwise there may be two separate tanks on a roof to store the two qualities of water separately.

6.4.2   *Combined Storage and Supply System*

When the main water or water from other independent sources is found to be unreliable for direct drinking then rainwater and water from those sources may be stored in the same tank. If the quality of mixed water is found suitable for direct use, particularly for nonpotable purposes, then the supply system can be developed by designing and installing one pumping and one piping system.

Furthermore, by treating such mixed water, the desired water quality can be obtained and by further disinfecting it, the water can be used as potable water. When a building must be supplied with potable water, it can be supplied by developing one pumping and piping system of treated and disinfected water and keeping it in a separate tank after treatment.

6.5 Sizing of the Storage Tank

Generally storage is the most expensive part of rainwater harvesting. Therefore, the sizing of the storage element should be calculated judiciously to avoid unnecessary expenses in construction as well as maintenance. Many other factors can affect the decision of sizing a tank to be used as follows:

1. Reliability on water to be availed from other sources
2. Expected demand of rainwater that must be stored for supplying
3. Restriction in the scale of the tank dimensions in relation to structural considerations
4. Availability of sufficient room to house the reservoir
5. Constructability
6. Aesthetics.

Various methods for sizing rainwater storage reservoirs exist. These methods vary in complexity and sophistication. Here, only the following two methods are described.

1. Demand-side approach (dry-season demand)
2. Supply-side approach (graphical methods).

6.5.1 Demand-Side Approach

This method is the simplest and most widely used one. Calculation of the storage requirement is based on the required water volume (consumption rates) and the occupancy of the building. This approach is well applicable in areas where a distinct dry season prevails. The reservoir is designed to meet the necessary water demand throughout the dry season. This method should be used in areas where there is sufficient rainfall as well as adequate catchment. This method helps in estimating tank size in the absence of rainfall data. To obtain the required storage volume following formula is used:

Demand = Water use per day × Population to be served × Number of dry days

$$(6.1)$$

Example:

Water consumption = 20 L/person (capita)/d
Population in the household = 5
Dry period (average dry period) = 150 days

Therefore, the rainwater-storage volume required = demand = 20 × 5 × 150 = 15,000 l.

The storage tank should be a little larger than the water volume to be stored for keeping free board above the stored rainwater surface.

6.5.2 Supply-Side Approach

This method is the most appropriate to estimate storage-tank size. In this method, rainfall data are required. The amount of available rainwater for storing depends on the amount of rainfall, the area of the available catchments, and the runoff coefficient of the catchments' surface.

The generation of rainwater from a catchment is usually represented by a runoff coefficient (RC). The runoff coefficient for any catchment is the ratio of the volume of water that runs off of a surface to the volume of rainfall that falls on the catchment area. A runoff coefficient of 0.8 of the catchment material means that 80 % of the rainfall can be collected from that catchment. Therefore, the higher the runoff coefficient, the more rainwater can be collected. An impermeable catchment will yield the highest volume of runoff. The runoff coefficient for various roofing materials is listed in Table 6.2.

Example:

Mean annual rainfall = 500 mm and a 5-month dry period of a season.
Building catchment area = 100 m^2
Runoff coefficient = 0.9.
Population = 10
Average water consumption = 20 L/person/d

Table 6.2 Runoff coefficient for flat catchments of various materials

Sl no	Type of materials	Runoff coefficient
1	GI sheets, metals, glass, slate	0.9–1.0 [7]
2	Glazed tiles	0.6–0.9 [3]
3	Clay tiles (hand made)	0.24–0.31 [4]
4	Clay tiles (machine made)	0.30–0.39 [4]
5	Cement tile	0.62–0.69 [4]
6	Aluminum sheets	0.8–0.9 [3]
7	Flat cement roof	0.6–0.7 [3]
8	Bituminous	0.7 [2]
9	Corrugated iron	0.8–0.85 [4]
10	Thatched	0.2 [3]
11	Asphalt fiberglass shingles	0.9 [5]
12	Plastic sheets	0.80–0.9 [6]
13	Terra cotta	0.70 [5]
14	Wood	0.65 [5]
15	Gravel	0.7 [7]

Source [2–7]

Therefore, water demand = 20 × 10 × 5 × 30 = 30,000 L.

Rainwater that could be supplied from the catchment = catchment area × rainfall × runoff coefficient = 100 m² × 300 mm × 0.9 = 27 m³ or 27,000 L/year.

Therefore, the potential annual rainwater availability is 27,000 L. There is shortfall in having 3000 L of rainwater/year. Under such conditions the storage-tank volume should be a little greater than 27,000 L.

6.5.3   Shape of Storage Tank

After finding the volume of rainwater to be stored, the volume of the storage tank must be determined, i.e., larger than the required volume of rainwater to be stored. If the determined tank volume falls within the commercially available prefabricated storage tanks, then a suitable sized tank, taking into account the placement of the tank, may be purchased from the market. If the storage tank must be constructed at the site, then the shape of the reservoir should be determined in accordance with the space available where it must be placed or constructed. In determining the shape of the constructed reservoir, the following steps should be followed.

1. The shape and the spatial area covered by the reservoir should be fixed depending on the shape of the area available for making the storage reservoir
2. The depth of rainwater storage can be determined by dividing the storage volume by the internal area of the storage tank determined.
3. If the depth found in step 2 is too deep or too shallow, then adjust the area of the reservoir. The usual choice of storage depth is approximately 1.5 m. The total depth of the storage tank should not be <1 m.
4. Add free board with the depth of storage. The depth of free board is considered to be between 150 and 300 mm to determine the total internal depth of the reservoir.

From a structural point of view, water reservoirs of a circular or curved shape is preferable to square or rectangular shaped ones, particularly when it must be built under the ground. If the depth of the reservoir goes too deep, then the cost of construction increases comparatively due to the increased cost of the relatively thick side walls and base. Again, in deep reservoirs, the area covered is less, so the cost of the top and bottom slabs of the reservoir would cost comparatively less. Therefore, the overall shape of the reservoir should be judiciously chosen for economy as well as safety.

Free board: Free board is the free height between the static water surface and the bottom surface of the top slab of the tank. In a water storage tank, free board of 150–300 mm is usually provided. In special cases, e.g., for avoiding slosh-dynamic

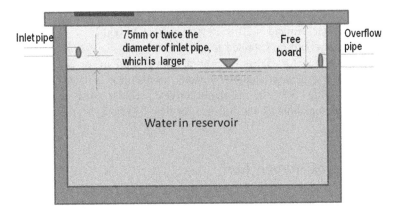

Fig. 6.7 Free board and related pipe positions in a tank

effects, the free-board height should be calculated. The bottom of inlet pipe of the tank should be at least 75 mm or twice the diameter of inlet pipe which is greater, above the static water level in the tank. The over flow pipe, should be one size larger than the inlet pipe size, the bottom of which should be at the static-water level as shown in Fig. 6.7.

6.6   Aesthetics

The size, material, surface finish, and location of a tank for storage, as proposed by the concerned engineers, may not be acceptable to the architect or building owner on aesthetic grounds. The "look" of a certain type and size of tank at a particular location may be demanded, by the architect, which may not be acceptable to the engineer from a structural point of view. In contrast, a proposal regarding tank by the engineer may not be acceptable to the architect from the perspective of aesthetics. From an aesthetic point of view, there are three broader aspects of disagreement among the decision-makers as follows:

1. Size
2. Shape and
3. Surface finish.

In case of disagreement regarding the size of the tank, there are three alternative approaches to come to a solution as follows.

1. Increase the height and decrease the area of the tank.
2. Decrease the height and increase the area of the tank.
3. Use multiple interconnected smaller tanks instead of one larger tank.

When there is any choice of making the reservoir a particular shape other than regular, that can also be performed; however, the following factors should be considered before making the decision:

1. Structural soundness
2. Constructability
3. Maintainability and
4. Cost.

Regarding the surface finish, the material used in constructing the reservoir governs the aesthetics. In some cases, the surface finish can be made of different materials other than the material that the reservoir is made of as follows:

1. Painting
2. Cladding by any surface finishing elements of different material
3. Walling by creepers (suitable for masonry reservoirs).

6.7   Functionality of a Storage Tank

To run the rainwater-harvesting system effectively, the storage tank must function well primarily to maintain the quality of rainwater up to the expected level. Therefore, all sorts of measures should be taken to prevent contamination of the stored rainwater. In addition to taking measures for preventing contamination, regular maintenance of the tank and safety of all components of the storage tank are other important issues to be well addressed for proper functioning. Tank material may also play important role in maintaining the quality of stored rainwater. For example, rainwater collected in plastic and plastic-lined water tanks remains naturally acidic and can corrode the copper pipes and fittings, thus resulting in leaching of copper into the supplied rainwater [8]. In contrast, storing rainwater in masonry tanks helps in neutralizing acidic rainwater to some extent.

Storage tanks also play an important role as a component of a rainwater-supply system depending on the type of main supply system in a building. For example, in an underground overhead-tank supply system, the roof-top tank becomes a facilitating component of the supply system rather than having an exclusively storage function.

For whatever purposes the tank is used, various measures must be taken to protect the stored water from pollution and contamination, allow easy maintenance of the tank, and provide safe use of rainwater. These measures are as follows:

1. The construction should ensure the water tightness of the reservoir.
2. The structural elements of the building that will remain in contact with the stored rainwater should be designed, constructed, and maintained as a water-retaining structural element.

3. Draw-off pipes on the storage tank should be at least 15 cm above the tank floor and not close to the inflow point.
4. Wire net shall must be provided to cover all inlets to prevent insects and mosquitoes from entering the tank.
5. The tank should be kept covered and light excluded to prevent growth of algae and microorganisms.
6. The tank must be placed in such a way so that the water temperature is kept as low as possible to limit bacterial and algae growth.
7. In case of potable rainwater storage, a backflow preventer in the form of either an air gap or a backflow assembly must be provided.
8. Tanks should be inspected and cleaned at least twice a year.
9. Sump should be made at a suitable location of the tank bottom.
10. The inner surface of the tank should be finished very smooth so as to have no cracks or holes.
11. The inner surface should be bright in colour. No toxic paint should be used.
12. Storage tanks should be provided with ventilating covers.
13. There should be two manholes located far apart.
14. Catch rings or a cats-ladder made of noncorroding material must be provided for deep reservoirs.

6.8 Prefabricated and Constructed Tanks

The use of prefabricated or commercially available water tanks is supposed to be more advantageous in various respects compared with constructed tanks of a similar size. However, there are also various disadvantages in using commercially available prefabricated tanks. Following is a comparison between these two categories of tanks.

6.8.1 Prefabricated Tanks

Advantages	Disadvantages
Can be installed quickly	No flexibility of choice regarding volume
Can be easily dismantled	Little flexibility in installing pipes of different diameter
Capacity can be changed by replacement	No provision of drainage pipe
Position can be easily shifted or relocated	Lifting at top of high-rise buildings is difficult.
Comparatively less weight	Additional pipe connection is needed for inter-connection of more tanks

6.8.2 Constructed Tanks

Disadvantages	Advantages
Construction takes considerable time	Flexibility of choice regarding volume
Dismantling and debris management is troublesome	Flexibility in shaping the reservoir befitting the site
Capacity cannot be increased easily	Flexibility in installing pipes of different diameter
Position cannot be easily shifted or relocated	Provision can be made for drainage pipe
Comparatively heavy weight	Lifting of construction materials is relatively easy
Leak-proof or water-tight construction may be challenging	

6.8.3 Sectional Tank

For storing a larger volume of rainwater, multiple prefabricated tanks may be inter-connected to meet storage requirements. In this case care, should be given to sizing the connection pipe among the tanks. Sufficient larger-diameter connection pipe should be used to maintain uniform filling of rainwater into the tanks.

Another way of storing a large volume of rainwater is using constructed sectional tanks with the help of prefabricated modular panels that are bolted together on-site [7]. These types of fabricated tanks are particularly useful because they enable some confined and irregular spaces inside buildings to be effectively used for rainwater storage.

References

1. International Code Council (ICC) (2015) Water Supply and Distribution. Chapter 6, Verginia Plumbing Code 2006, p 6-3. https://www2.iccsafe.org/states/Virginia/Plumbing/PDFs/Chapter%206_Water%20Supply%20and%20Distribution.pdf. Retrieved on 25 Nov 2015
2. Ward S, Memon FA, Butler D (2010) Harvested rainwater quality: the importance of appropriate design. Water Sci Technol 61(7):1707–1714. Cited in Farreny R, Morales-Pinzón T, Guisasola A, Tayà C, Rieradevall J, Gabarrell X (2011) Roof selection for rainwater harvesting: quantity and quality assessments in Spain. Water Res 45(2011):3245–3254. http://icta.uab.cat/ecotech/2011_Roof_selection_for_rainwater.pdf. Retrieved on 20 Jan 2016
3. Worm J, van Hattum T (2006) Rainwater harvesting for domestic use. Agromisa Foundation and CTA, Wageningen, The Netherlands
4. Zhu Q, Liu C (1998) Cited in Rainwater harvesting for agriculture and water supply by Zhu Q, Gould J, Li Y, Springer, p 243

5. Rainwater Harvesting Community (2012) Roof materials. https://www.facebook.com/rainwater2sustain/posts/354371111318757. Retrieved on 24 Jan 2016

6. Khoury-Nolde N (2016) Rainwater harvesting. http://www.rainwaterconference.org/uploads/media/Rainwater_Harvesting_-_an_overview_.pdf. Retrieved on 24 Jan 2016

7. Grundfos (2015) Rainwater harvesting in commercial buildings. Application guide, p 17. https://www.grundfos.com/content/dam/Global%20Site/campaigns/Grundfos-isolutions/5/download-pdfs/7-CBS-RWH-appl-guide_LowRes.pdf. Retrieved on 26 Nov 2015

8. ABC Southeast NSW (2010) Copper poisoning linked to plastic water tanks by Bill Brown. http://www.abc.net.au/local/stories/2010/11/24/3075218.htm. Retrieved on 1 Oct 2015

Chapter 7
Rainwater Conditioning

Abstract Rainwater is pure when formed in the cloud, but it becomes contaminated or characteristically changed while falling and flowing over any surface due to absorbing various suspended elements present in the air, and the dirt or chemicals present on the surfaces, coming into contact with the rain or rainwater. As a result, it is rarely possible to obtain pure rainwater for direct consumption, particularly water that is collected in buildings of urban areas. Storing rainwater for a long period might cause further contamination due to the growth of various microorganisms in the reservoir. Therefore, after collecting and storing it, it is necessary to purify contaminated rainwater according to the purpose of use. There are various methods of purifying or conditioning water and thus rainwater. In buildings, considering the qualitative and quantitative aspects of water demand, limited methods of treatment and conditioning systems are generally employed in rainwater harvesting. The treatment methodologies generally adopted are screening, sedimentation, filtration, and disinfection. In this chapter, the usually practiced treatment process recommended for different purposes of uses of rainwater in buildings, are discussed. At the end of the chapter, the planning aspects of treatment systems are delineated.

7.1 Introduction

It has already been discussed that rainwater will contain impurities received from various sources during its falling and flowing on various surfaces before reaching the collecting storage tank. Therefore, conditioning or treatment of collected rainwater is almost inevitable in whatever form or scale it may be. Considering the purpose of using rainwater, its quality at the end point should be ensured accordingly, for which a particular or a combination of various conditioning or treatment is needed. In this chapter, various aspects of rainwater-conditioning and -treatment processes are discussed.

© Springer International Publishing Switzerland 2017
S.A. Haq, PEng, *Harvesting Rainwater from Buildings*,
DOI 10.1007/978-3-319-46362-9_7

7.2 Methods of Conditioning Rainwater

There are various ways of conditioning rainwater. The system adopted should be based on the purpose of the rainwater use. Therefore, on the basis of the method of conditioning requirement, the purposes of rainwater uses can be grouped in the following categories:

1. For drinking, cooking, washing utensils, bathing, and similar other uses
2. For clothes washing, floor washing, fountain, waterfall cascade, and similar other uses
3. For use in fire fighting, air conditioning, and similar other uses
4. For toilet flushing, gardening, cleaning artificial grounds and parking lots, and similar other uses.

Rainwater should be treated adopting the following various methods corresponding to the purpose of use mentioned below:

1. For using rainwater in drinking, cooking, washing utensils, bathing, ablution, and any other uses related to culinary, eating, and body-contact activities, it should be filtered and disinfected.
2. For clothes washing, floor washing, fountain, waterfall cascade, and any other sanitation and recreational purposes, rainwater should be filtered.
3. For use in fire fighting, air conditioning, and other utility services, sedimentation of suspended particles in rainwater is required.
4. For toilet flushing, cleaning artificial grounds and parking lots, etc., screening for collecting floating materials is needed.
5. For plant watering or gardening, no treatment is needed [1]. Acidic rainwater must be avoided for these purposes.

From the above discussion it is made clear that for the conditioning or treatment of rainwater, one or a combination of the following treatment methodologies is necessary.

1. Screening
2. Sedimentation
3. Filtration
4. Disinfection.

7.3 Screening

Rainwater from any catchments may contain significant quantities of plant debris, leaves, papers, and many other solids that remain mostly suspended or floating and should be removed first. Screens are used in rainwater conditioning to strain those floating and suspended matters or particles from the storm or rainwater stream. Screening is the passing of water through small bars or wire mesh in order to retain any matter whose size is greater than the size of the opening between the bars or mesh of the screen. Screening is usually introduced as the first step in the rainwater-treatment system, and screens are the first component used for this purpose. Screening retards the flow, so its mesh size should be selected properly. Screens should be placed in such location so that they can be easily accessible and cleaned easily. Uninterrupted flow must be ensured through the regular cleaning of screens.

Depending on the opening size, screens are classified as below:

1. Bar screens
2. Coarse screens
3. Fine screens
4. Micro-screens.

The opening of bar screens is >25.4 mm; coarse screens have clear openings of 4.8–25.4 mm; fine screens have openings of 0.1–4.8 mm; and micro-screens have clear openings <0.1 mm [2]. If sands are to be removed from the rainwater, a screen with 300-mesh size can be used [3], but the use of such screen in rainwater-collection pipe greatly retards the collection and should be cleaned quarterly [3].

In flat catchments, the dome-shaped grating, as shown in Fig. 5.19 in Chap. 5, placed on the inlet of rainwater down pipes serves the purposes of screening. On gutters, coarse screening mesh can be provided as shown in Fig. 7.1. Fine screening is suggested to be used at the overflow outlet of the storage tank to act as a barrier to mosquito or any other vectors entering the reservoir [4].

Screens are to be maintained properly to ensure uninterrupted collection of rainwater. Regular monitoring of screens and cleaning is to be performed, particularly during the collection period, and before the rain starts. To reduce the accumulation of matters on the screen, the catchments should be well maintained.

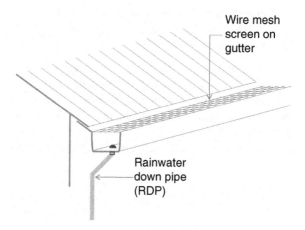

Fig. 7.1 Wire-mesh coarse screen installed on a gutter under a sloped roof

Wire mesh screen on gutter

Rainwater down pipe (RDP)

7.4 Sedimentation

During the process of collection, very small and fine suspended solids present in the rainwater are allowed to settle by gravity under the almost-still conditions of the collected rainwater. The sedimentation process is based on the gravitational settling of discrete particles. During the process of settling, particles that do not change in size, shape, or mass are known as "discrete" particles. In rainwater conditioning, a nonflocculent, free-settling type 1, plain sedimentation process is generally employed. The incorporation of plain sedimentation has the following advantages:

1. Plain sedimentation lightens the load on the subsequent treatment process.
2. The operation of subsequent purification process can be better controlled.
3. The cost of cleaning the chemical coagulation basins is decreased.
4. No chemical is lost with sludge discharged from the plain settling basin.
5. The types and quantity of chemicals used in subsequent treatment processes are reduced or eliminated.

7.4.1 Design Aspects of Sedimentation Tanks

The design aspects of sedimentation tanks are as follows:

1. Velocity of flow
2. Capacity of tank
3. Inlet and outlet arrangement
4. Shape of tank
5. Miscellaneous considerations.

Velocity of flow: The velocity of flow of rainwater in sedimentation tanks should be sufficient to cause the hydraulic subsidence of suspended impurities. The amount of matter removed by a sedimentation tank depends on the following factors.

1. Velocity of flow in the sedimentation tank
2. Size and shape of particles
3. Viscosity of rainwater.

In a sedimentation tank, the percentage of particle removal is dependent on the settling velocity V_s of the particles and the overflow rate V_o. The trajectory of a settling model of particles in a sedimentation tank is shown in Fig. 7.2. When the settling velocity is equal to the overflow rate, then 100 % removal of suspended particles can be achieved.

The percentage of particle removed P can be expressed as follows:

$$P = 100\frac{V_s}{V_o} \tag{7.1}$$

Fig. 7.2 Model of the
trajectory of settling particles
in a sedimentation tank

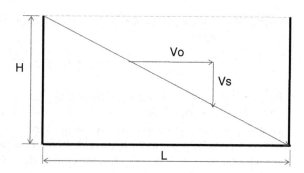

where V_s = settling velocity; and V_o = overflow rate.

The settling velocity V_s is the uniform velocity of a settling discrete particle in a quiescent fluid contained in a settling tank. This velocity can be described by Stroke's Law as shown below.

$$\text{The settling velocity } V_s = \frac{g}{18\mu}(\rho_s - \rho)d^2 \text{ cm/s} \tag{7.2}$$

where ρ_s is the mass density of particles; ρ is the mass density of rainwater, which is considered to be 1; d is the diameter of particles (mm); μ is the viscosity of rainwater (cm^2/s); and g = is the acceleration due to gravity (981 cm/s^2).

Smaller particles have lower settling velocities. Throughout the tank, the settling velocity should remain uniform. Generally the settling velocity is not allowed to exceed the range of 0.25–0.5 cm/s.

Overflow rate V_o, also called "surface loading," is the flow of water per unit surface area of the tank. It can be expressed as cubic metres per square metre of surface area of the tank per unit of time. It is assumed that the settlement of a particle at the bottom of the tank does not depend on the depth of the tank but actually depends on its surface area.

Surface overflow rate,

$$V_o = \frac{Q}{L \times B} \tag{7.3}$$

where Q is the discharge or rate of flow; L is the length of the tank; and B is the breadth of the tank.

Example

A sedimentation tank has an overflow rate of 0.2 mm/s. Particles present in the rainwater have settling velocity of 0.1 mm/s.

Therefore, the percent removal $P = 100 \times 0.1/0.2 = 50$ %.

Capacity of tank: The capacity of sedimentation tank C is calculated by the following formula:

$$C = Q \times T \qquad (7.4)$$

where Q is the discharge or rate of flow; and T is the detention period.

The detention period of a settling tank is the theoretical time for which the incoming rainwater remains detained in the tank. Virtually the theoretical time taken by a particle in rainwater to travel between entry and exit of a settling tank is known as the "detention period." It is the ratio of the volume of the basin to the rate of flow through the basin.

$$\text{Detention period } T = \frac{\text{Distance of descend}}{\text{Velocity of descend}} \text{ i.e., } T = \frac{H}{V_s} \qquad (7.5)$$

where H is the liquid depth of the sedimentation tank; and V_s is the settling velocity.

The detention period depends on the size of the suspended impurities present in rainwater. For plain sedimentation tanks, the detention period varies from 3 to 4 h.

Inlet and outlet arrangements: The inlet should be devised in such a way that it provides uniform distribution and velocity of inflowing rainwater inside the tank. A baffle is usually constructed across the basin, in front and close to the inlet, which projects to the middle of the tank to dissipate inlet velocity and provide uniform flow. An outlet weir or submerged orifice is constructed behind the outlet of the tank to maintain velocity of flow suitable for settling in the basin and to minimize short-circuiting. These arrangements should be properly designed and located in such a way that they do not form any obstruction or cause any disturbance to the flowing rainwater. For pumping rainwater from sedimentation tank special measures must be taken around suction inlet, as shown in Figs. 3.1 and 3.4 of Chap. 3, so that sediments are not disturbed.

Shaping of tanks: Sedimentation tanks may be designed in three different shapes with respect to three different types of flow created in the tank. Among three types of tanks, a rectangular-type sedimentation tank, as shown in Fig. 7.3, is mostly preferred in treating rainwater. In this type of tank, the flow is maintained horizontal. Rectangular tanks of narrow width, shallow depth, and long length have flow stability and minimize short-circuiting. For satisfactory performance, the sedimentation tank should be designed and shaped according to the parameters listed in Table 7.1.

7.5 Filtration

Filtration is a mechanical or physical operation performed in water treatment primarily for the separation of suspended solids from water by incorporating a medium, called a filter medium, through which the water can pass but the solids are retained for subsequent removal. The filter medium virtually blocks the passage of

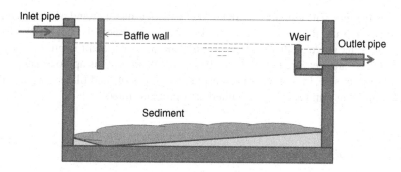

Fig. 7.3 Rectangular sedimentation tank

Table 7.1 Parameter for designing and shaping the sedimentation tank

Sl. No.	Parameters	Value
1.	Detention period	1.5 [5] to 5 [6] h
2.	Overflow rate	12–18 m³/d/m² of tank [7]
3.	Velocity of flow	Not more than 30 cm/min [7]
4.	Length-to-breadth ratio	3:1[8] to 5:1 [5, 8]
5.	Depth of tank	2.5–5.0 m (3 m) [7]
6.	Weir loading	300 m³/m/d [9]
7.	Solids removal efficiency	50 % [9]
8.	Turbidity of rainwater after sedimentation	15–20 NTU [9]
9.	Inlet and outlet zones	0.75–1.0 m [9]
10.	Free board	0.5 m [9]
11.	Sediment zone	0.5 m [9]
12.	Bottom slope	100:1 toward inlet [7]

Source [5–9]

suspended matter in flowing water through physical obstruction, chemical adsorption, or a combination of both. In rainwater harvesting, filtration not only removes suspended matters, it also removes microorganisms present in the rainwater. Therefore, filtering removes both suspended and faecal matters.

The efficiency of filtration is mainly dependent on the size and depth of the filtering medium. It has been observed that the percent removal of suspended solids and fecal coliforms through sand filtering can be increased by increasing the depth of the filtering medium [10].

For rainwater harvesting in a building, generally sand filters are used. The filtering apparatus basically contains sand and gravel as filtering media. The sand virtually does the filtering, and the gravel keeps the sand from getting out of position. There is an under-drain, through which the filtered rainwater flows out. After the filter is operated for a certain period, the sand becomes clogged with retained particles. Then the filter must be backwashed by reversing the flow through the filter medium The clogging particles, being lighter than the sand, rise up and are flushed away from the system. When backwashing is stopped, the sand settles back down onto the gravel. Filtering is then run again as usual, and the process begins anew. Filters are measured in microns. For comparison, sand is approximately 100 to 1000 μm in size. In a broader perspective, two types of sand filters are generally used in treating rainwater: gravity filters and pressure filters.

7.5.1 Gravity Filters

In this process, filtration is performed by applying gravitational flow of water or rainwater. Gravity filters are again of two types: slow sand filters and rapid sand filters. These filter beds require minimum maintenance, except for periodic scraping of the fine clay and silt deposited on the filter bed. Silt deposited on the filter medium should be cleaned regularly by removing the top deposited silt. Once per year, the top 5–10 cm sand layer should also be scraped to maintain a constant flow rate through the filter material.

Slow sand filter: Slow sand filters, which can remove bacteria and very small particles, are best suited for the filtration of rainwater to be used for drinking, cooking, etc. The sand used for the filtration is specified by the effective size and uniformity coefficient of the sand used. The effective size D_{10} is the sieve size, in millimeters, that permits 10 % sand by weight to pass through. The uniformity coefficient of filtering sand can be calculated by the ratio D_{60}/D_{10}.

A slow sand filter is made up of two distinct layers: a top filtering layer of fine sands and a drainage layer of coarse sands at the bottom. In the top layer, the fine sands are of effective size 0.15–0.35 mm, which preferably should have uniformity coefficient of 1.5 to <3.0 [11]. The thickness of the filter bed layer may be 0.6–1.2 m [12]. Under the filter bed layer, there are drainage layers of three grades of coarse sand of varying sizes ranging 2–8, 8–16, and 16–32 mm [13]. The voids created by these graded layers of coarse sand do not permit filtered fine sands to pass through it. The thickness of these layers may be varying from 400 to 600 mm maximum [14]. The coarse sand is laid in layers in such a way that the smallest sizes are placed at the top and with the sizes gradually increasing downward. The water level on the filter medium, which acts as head for flow-through filter media, is maintained 1.0–1.5 m [12]. Filtered water is collected evenly from beneath the filter media by a properly designed under-drain piping system as shown in Fig. 7.4.

The gravity filtration rate of a slow sand filter is usually found to range from 100 to 200 L/h/m^2 surface area of filter bed [13]. For this type of filter, a comparatively larger area is required. Turbidity >30 NTU in the rainwater can cause the filter to clog rapidly [15]. The filtration rate becomes slow as the growth of biofilm around the sand particles thickens. To refurbish the filtering performance two methods are applied: (1) The top few millimetres of fine sand, as mentioned earlier, is scraped off and replaced by new layer of clean sand; and (2) the filter bed is backwashed when replacing of the top sand layer does not considerably improve the performance.

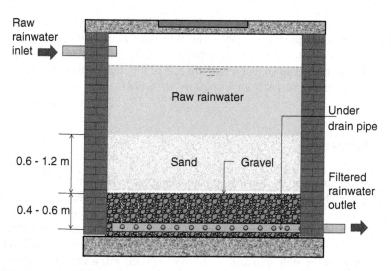

Fig. 7.4 Slow sand filter

Rapid sand filter: For rapid filtration, there must be a comparatively high velocity of flow through the filtering medium. A rapid sand filter exhibits a high rate of filtration capacity compared with that of slow sand filtration. The filtration rate of rapid sand filters ranges from 4000 to 12,000 L/h/m^2 [16]. In rapid sand filtration, much coarser sand is used and thus a comparatively small area of filter bed is required. Rapid sand filtration is preferred when rainwater is to be used for ground recharging as well as for uses other than drinking or cooking.

Two layers of media are provided in rapid sand filtration. The top layer is virtually the filter media comprising comparatively coarse sand having an effective size of 0.35–1.0 mm and a uniformity coefficient ranging from 1.2–1.7 [11]. The depth of this layer is made 60–80 cm. [11]. The bottom layer is a supporting medium made of gravel of size varying 2.5–50 mm, keeping the increased size toward the bottom, and the thickness is made 45–50 cm [17]. Under the gravel layer, there is an under-drain piping system, made of perforated pipes at the bottom half, by which filtered water is drawn out and backwash water is pushed into.

Rapid sand filters used to clog comparatively at shorter time intervals due to their high rate of filtration. This type of filters is usually cleaned at regular intervals of every 24–72 h, but sometimes they are cleaned every few hours [16].

7.5.2   Pressure Filters

A pressure filter is one kind of rapid sand filter, built in a closed water-tight cylindrical drum, in which the water to be filtered passes through the sand bed under pressure greater than the atmospheric pressure. This type of filter is fitted with pumps to pressurize the water through the filter chamber. Complete units of pressure filters are commercially available. The main disadvantage of these filters is that they require energy for operation, plus they need to be backwashed periodically to remove the finer material so that a uniform rate of filtration is maintained. The rate of filtration normally varies from 7 to 20 $m^3/m^2/h$ [18].

7.6   Disinfection

Infectious diseases can be caused by pathogenic bacteria, viruses, and protozoan parasites, which might be present in rainwater. Some waterborne pathogenic microorganisms may spread by rainwater and can cause severe, life-threatening diseases such as typhoid fever, cholera, and hepatitis A or E, etc. Other microorganisms induce less dangerous diseases. Human beings may be introduced to these microorganisms through drinking such contaminated rainwater and suffer from these diseases.

Disinfection of rainwater should be performed for the removal, deactivation, or killing of pathogenic microorganisms present in collected rainwater. By the process of disinfection, microorganisms are destroyed or deactivated by termination of growth and reproduction.

Testing rainwater for microbial contamination: Before undergoing disinfection, rainwater should be examined for microbiological contamination by undergoing various optional tests such as thermo-tolerant coliform count (also known as "faecal coliform count," *Escherichia coli* (*E. coli*) count, or the simple H_2S test. Other traditional test methods are the most probable number (MPN) and membrane-filtration methods. According to Faisst and Fujioka, the H_2S test has been found to correlate well with thermo-tolerant coliform levels in rainwater tanks [19].

Disinfection methodologies: Disinfection is performed by means of adopting some physical or chemical methods. Not all of the methods of disinfection of water may be suitable for disinfecting rainwater for harvesting in buildings. Suitable methods of disinfection, i.e., those that can be performed in rainwater harvested in buildings, are as follows:

A. Physical disinfectants and the corresponding methods

1. Heat by boiling
2. Ultraviolet light (UV)

B. Chemical disinfectants and methods

1. Chlorine (Cl_2) by chlorination
2. Ozone (O_3) by ozonation

7.6.1 Boiling

Boiling rainwater can be an effective method of disinfection because no important water-borne diseases are caused by heat-resistant organisms. A container of rainwater is placed over a source of heat sufficient to raise the temperature to the boiling point of water. The temperature is then increased to the boiling point. At sea level, the boiling point of water is 100 °C or 212 °F at 1 atmosphere of pressure. The rainwater should be boiled at this temperature for approximately 1 min [20]. Although some highly resistant organisms may survive after boiling for this length of time, it is believed that such presence is rare and so is acceptable [20]. To make boiling most effective, rainwater should be boiled for at least 20 min [21].

The boiling point of water, also of rainwater, depends on the atmospheric pressure, which changes according to elevation. One additional minute of boiling is required for every 330 m above sea level to account for the decrease in boiling temperature at different altitude. The boiling point of water also depends on the purity of the rainwater. Rainwater that contains impurities should be boiled at a higher temperature than comparatively pure rainwater. Because of the inconvenience, boiling is not generally performed to treat huge volumes of rainwater except in emergencies. For heating a large quantity of rainwater, a boiler may be used. The heating times and temperatures required to make drinking rainwater safe from different pathogens is listed in Table 7.2. In all cases, placing rainwater in a rolling boiler for 2 min ensures that rainwater is safe from pathogens.

The advantages of boiling rainwater include the followings:

1. Boiling easily makes rainwater safe to drink when it is contaminated with living organisms.
2. Pathogens in rainwater are killed if the rainwater is boiled for a sufficiently long time.
3. Boiling will expedite the driving out of some volatile organic compounds (VOC) that might be present in rainwater.

The disadvantages of boiling rainwater include the following:

Table 7.2 Rainwater-heating temperatures and times for pathogens

Organisms	Pathogen	Temperature (°C)	Inactivation time (s)
Bacteria	E. coli	65	<2
	Legionella pneumophila	58	360
	Salmonella choleraesuisa	60	300
	Shigella sonnei	65	3
	Vibrio cholerae	70	120
Virus	Hepatitis A	75	30
Protozoa	Cryptosporidium parvum	60	300
	Giardia	56	600

Source [22]

1. Rainwater must not be boiled when it is contaminated due to the presence of toxic metals, chemicals (e.g., lead, mercury, asbestos, pesticides, solvents, etc.), or nitrates in considerable amounts.
2. Boiling rainwater for a long period may cause a higher concentration of many harmful inorganic contaminants that do not vaporize off while the rainwater does evaporate.
3. Energy is needed to boil the rainwater, so it may not be possible or may be difficult in an emergency situation.

7.6.2 Ultraviolet Light

The use of ultraviolet (UV) light is an attempt to imitate nature. An ultraviolet disinfection system can be employed for disinfecting rainwater to be used in buildings of various occupancies because it is simple and easy to handle. UV units, as shown in Fig. 7.5, can be installed easily on a rainwater harvesting system, generally in reservoirs, in order to safeguard rainwater quality.

Exposing rainwater to ultraviolet light destroys the pathogens that are present. To assure thorough treatment, the rainwater must be free of turbidity. Otherwise some bacteria are protected from the germ-killing ultraviolet rays. For effectiveness and assurance, it must be closely checked that sufficient ultraviolet energy is reaching all of the points of application at all times in the tank. UV is typically used as a final disinfection stage after filtration. Otherwise suspended solids present in the water can effectively shield pathogens from UV light, which may not be destroyed.

The advantages of ultraviolet light include the following:

1. It is an automatic disinfection system.
2. No known toxic or nontoxic byproducts are used.

Fig. 7.5 Schematic diagram of a UV disinfection unit

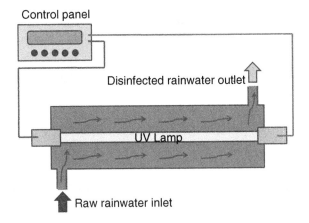

3. There is no chance of creating a bad taste or odour. Rather it improves the taste of rainwater by killing nuisance microorganisms and destroying organic contaminants.
4. There is low contact time.
5. Treatment has no residual effect.
6. There is no danger of overdosing.
7. It does not affect minerals in the water.
8. All types of pathogenic microorganisms are either killed or rendered inactive.

The Disadvantages of ultraviolet light include the following:

1. It has low penetration power, and its effectiveness is shielded by turbidity.
2. A slime layer develops on the UV light tube.
3. There is no simple test of the results.
4. It requires electricity to operate. When the power is off, purification will not occur.
5. The ultraviolet tube gradually loses power (see http://www.freedrinkingwater. com/water-education2/82-form-lower-organisms.htm).
6. UV radiation may not be effective for rainwater containing high levels of suspended solids, soluble organic matter, turbidity, or colour.
7. Most UV units are usually insensitive to temperature and pH differences in rainwater.
8. No standardized mechanism measures, calibrates, or certifies how well the equipment works before or after installation.
9. UV light is not effective in destroying nonliving elements, chemical contaminants (e.g., lead, asbestos, chlorine, etc, and organic chemicals.
10. UV light is not effective in destroying tough cryptosporidia cysts that might be present in the rainwater.
11. UV light can further degrade the rainwater quality by converting some organic compounds into equally harmful byproducts.
12. UV is more effective against parasites than against viruses.

7.6.3 Chlorination

Chlorine acts as a good disinfectant in making rainwater potable by disinfection. Chlorine causes deactivation of most microorganisms present, and it is relatively cheap. Chlorine can easily be applied, measured, and controlled in the process of disinfection of rainwater. Adding chlorine to rainwater also results in the formation of a variety of chloro-compounds; the reactions lower the pH, thus making the rainwater more corrosive and aggressive.

Chlorination provides some residual chlorine in the stored and supplied rainwater, which is desirable to decrease the chance of regrowth of pathogens. According to WHO guidelines, the free residual chlorine concentration in drinking water, and thus in rainwater, should be between 0.2 and 0.5 mg/L [23]. The maximum chlorine concentration allowable in rainwater for disinfection is 5 mg/L [24].

In disinfecting rainwater, chlorine is seldom directly applied to rainwater to be treated. As a source of chlorine, bleaching powder (calcium hypochlorite: $CaOCl_2$) and commercial bleach solution (sodium hypochlorite: $NaClO$) are used. These are typically sold at certain concentration of "available chlorine." With the available concentration of chlorine desired, a concentration of chlorine solution is prepared using the following formula:

$$C_1 \times V_1 = C_2 \times V_2 \tag{7.6}$$

where C_1 is the available concentration of chlorine in volume V_1; and C_2 is the desired concentration of chlorine in desired volume V_2.

Example:

As a general guide, chlorine doses of approximately 5 mg/l can be produced by adding 40 mL of liquid sodium hypochlorite (12.5 % available chlorine) per 1000 L of water or 7 g of powdered calcium hypochlorite (75 % available chlorine) per 1000 L of water [25].

A commercial bleach solution of sodium hypochlorite ($NaClO$) is typically sold at a concentration of 5.25–12.5 % available chlorine. To obtain 5 ppm chlorine (Cl) solution, the volume of bleach solution, having 12.5 % available chlorine, which should be added to 1000 L of water, can be found in the following way:

From Eq. 7.6, we obtain $V_2 = C_1 \times V_1/C_2$.

12.5 %   available   chlorine = 12.5 g NaClO/100 mL =   125,000 mg NaClO/l or ppm

Therefore, the volume of bleach solution required = 5 × 1000/125,000 = 0.04 L, say, 40 mL.

Therefore, 40 mL of standard sodium bleach solution of 12.5 % available chlorine concentration should be added to the 1000 L of water in order to obtain a 5-ppm chlorine concentration in the tank.

Calcium hypochlorite contains 70–75 % available chlorine. To make a 5-ppm chlorine concentration, the volume of bleach powder that should be added to 1000 L of water is approximately 7 g. The amounts of chlorine, from various sources of a particular concentration of available chlorine, required to disinfect rainwater safely is listed in Table 7.3.

Table 7.3 The amounts of chlorine, from various sources of particular concentrations of available chlorine, required to disinfect rainwater

Treatment	Calcium hypochlorite 60–70 %	Sodium hypochlorite 12.5 %
Initial dose	7 g/1000 L	40 mL/1000 L
Weekly	1 g/1000 L	4 mL/1000 L

Source [26]

Disinfection of rainwater must be performed after the construction of, the pipe installation in, and every repair work performed on the tank or reservoir. To disinfect stored rainwater in the tank, the tank should be cleaned first. The tank is then filled partly with rainwater up to a certain level. Then the desired amount of chlorine-generating solution or powder is added to the water that will produce a chlorine concentration of 5 mg/L in the full tank. The tank is then filled with rainwater up to its overflow level. After at least 30 min of contact time at a pH value ≤8, the test for residual free chlorine should be performed. The concentration of residual free chlorine must not exceed 0.5 mg/L.

Stabilized chlorine (chlorinated cyanurates) should not be used for chlorination. The following guidelines should be followed in using chlorine.

1. Water cannot be poured onto chlorine, but chlorine can be added to water.
2. Skin contact with chlorine should be avoided.
3. Chlorine should be stored in a cool, dark place and out of reach of children.
4. Chlorine solutions should not be kept standing or exposed to air or sunlight for a long period.

7.6.4 Ozonation

Ozone is a tri-atomic form of oxygen (O_3) that is the strongest oxidant among the common disinfecting agents. Ozone is also used to control the colour and taste of water. Ozone is unstable, and it cannot be produced at, transported to, and thus generated at the point of use. Ozone can be produced by an ozone generator applying either electrical the corona-discharge principle, which is common, or ultraviolet irradiation of dry air or oxygen. Ozone is either injected or diffused into the stream of rainwater.

Advantages of ozonation

1. Ozone is extremely active as a disinfectant.
2. Ozone does not produce potential harmful byproducts and residuals.
3. A wider range of organisms is killed by ozonation than by chlorination.
4. Ozone results in excellent removal of taste and odors.
5. The reactions of ozone are more rapid than those of chlorination processes.

Disadvantages of ozonation

1. Ozonation may not kill cysts and some other large organisms.
2. Ozone must be generated before use, and the operating costs can be quite high.
3. Ozone has active residual. Therefore, its use in large distribution systems is not advised.
4. Ozone gas is poisonous, so proper ventilation is required.

7.7   Planning the Conditioning System

During the planning of spaces for various uses in building, the conditioning system must be brought in mind and, most importantly, the conditioning options to be adopted must be considered. Disinfection may not require any extra floor area to accommodate any disinfecting appliances. Filtration system needs some area to house the filtration unit and a tank to accumulate the filtered water. Planning is also needed to place the filtering units. In gravity filtration, the units are placed one below the other. This filtration system is economical and therefore area should be allocated at different levels.

References

1. Group Raindrops (1995) Rainwater and you 100 ways to use rainwater. In: Tokyo international rainwater utilization conference, Japan, p 120
2. Pfafflin JR, Ziegler EN (2006) Encyclopedia of environmental science and engineering, 5th ed, vol 2 M–Z. CRC Press, Taylor and Francis Group, London, p 1191
3. Pushard D (2015), HarvestH2o potable rainwater: filtration and purification. http://www.harvesth2o.com/filtration_purification.shtml#.VgpN2dKqqko. Retrieved on 29 Sept 2015
4. The Caribbean Environmental Health Institute (2009) Rainwater: catch it while you can. In: A handbook on rainwater harvesting in the Carribean, p 26
5. Tiffany B (1991) Wastewater engineering: treatment, disposal, and reuse, 3rd ed. Metcalf and Eddy (Cited in Technical memorandum). http://www.kingcounty.gov/ ~ /media/services/environment/wastewater/industrial-waste/docs/TechAssistance/CDW_SedTank_Tech_Memo1111.ashx?la=en. Retrieved on 06 June 2016
6. Caltrans (2001) Field guide to construction site dewatering—Appendix B: sediment treatment options. California State Department of Transportation (Caltrans)—Construction Division. Publication No. CTSW-RT-01-010. October 2001 (Cited in technical memorandum, Contributed by Bruce Tiffany) http://www.kingcounty.gov/ ~ /media/services/environment/wastewater/industrial-waste/docs/TechAssistance/CDW_SedTank_Tech_Memo1111.ashx?la=en. Retrieved on 06 June 2016
7. Indian Institute of Technology (2006) Water and wastewater engineering. http://nptel.ac.in/courses/105104102/Lecture%206.htm. Retrieved on 25 Jan 2016
8. Michael R (1997) Civil engineering manual for the PE exam, 6th ed (Cited in 'technical memorandum' Contributed by Bruce Tiffany). Lindeburg. http://www.kingcounty.gov/~/media/services/environment/wastewater/industrial-waste/docs/TechAssistance/CDW_SedTank_Tech_Memo1111.ashx?la=en. Retrieved on 06 June 2016
9. Venkateswara Rao P (2005) In: Srinivasa Rao K (ed) Water supply engineering. State Institute of Vocational Education, Hyderabad, p 45. http://bie.telangana.gov.in/Pdf/WaterSupplyEngg.pdf. Retrieved on 07 June 2016
10. Nassar AM, Hajjaj K (2013) Purification of storm water using sand filter. J Water Res Prot 5:1007–1012. file:///C:/Users/User/Downloads/JWARP_2013110513233519.pdf. Retrieved on 25 Jan 2016
11. United Nations Environment Programme (1997) Source book of alternative technologies for freshwater augmentation in Latin America and the Caribbean. Filtration systems. http://www.oas.org/DSD/publications/Unit/oea59e/ch24.htm. Retrieved on 1 Oct 2015
12. World Health Organization (1974) Slow sand filtration, pp 18–19. http://www.who.int/water_sanitation_health/publications/ssf9241540370.pdf?ua=1. Retrieved on 29 Sept 2015

13. Ludwig A (2015) Slow sand filtration. http://oasisdesign.net/water/treatment/slowsandfilter. htm. Retrieved on 29 Sept 2015

14. Qasim SR, Motley EM, Zhu G (2002) Water works engineering' planning design and operation. Prentice-Hall of India Pvt. Ltd, New Delhi 366

15. Brikke F, Bredero M (2003) Linking technology choice with operation and maintenance in the context of community water supply and sanitation. A reference document for planners and project staff. World Health Organization and IRC Water and Sanitation Centre, Geneva. http://www.who.int/water_sanitation_health/hygiene/om/wsh9241562153.pdf. Retrieved on 30 Sept 2015

16. World Health Organization (WHO) (1994) Fact sheets on environmental sanitation, 2.14 Rapid Sand Filtration, Geneva. http://www.who.int/water_sanitation_health/hygiene/emergencies/fs2_14.pdf?ua=1. Retrieved on 1 Oct 2015

17. The Water Treatment (2014), Rapid sand filters. http://www.thewatertreatments.com/water-treatment-filtration/rapid-sand-filters/. Retrieved on 1 Oct 2015

18. Delft University of Technology (2015) Water treatment, granular filtration, OpenCourseWare educational resources. http://ocw.tudelft.nl/fileadmin/ocw/courses/DrinkingWaterTreatment1/res00067/embedded/!46492046696c74726174696f6e32303037.pdf. Retrieved on 1 Oct 2015

19. Faisst EW, Fujioka RS (1994) Assessment of four rainwater catchment designs on cistern water quality. In: Proceedings of the 6th international conference on rainwater catchment systems, Nairobi, Kenya (Cited in 'rainwater harvesting and health aspects working on WHO guidance' Namrata Pathak and Han Heijnen). http://www.ctahr.hawaii.edu/hawaiirain/Library/papers/Pathak_Namrata.pdf. Retrieved on 1 Oct 2015

20. United States Environmental Protection Agency (1998) Emergency disinfection of drinking water—boiling, memorandum. http://nepis.epa.gov. Retrieved on 26 Jan 2016

21. Organization of American States (OAS) (2016) Disinfection by boiling and chlorination. https://www.oas.org/dsd/publications/Unit/oea59e/ch23.htm. Retrieved on 27 Jan 2016

22. World Health Organization (WHO) (2015) Boil water. Technical Brief. http://www.who.int/water_sanitation_health/dwq/Boiling_water_01_15.pdf. Retrieved on 07 Oct 2015

23. World Health Organization (WHO) (2011) Measuring chlorine levels in water supplies. Technical notes on drinking-water, sanitation and hygiene in emergencies. http://www.who.int/water_sanitation_health/publications/2011/tn11_chlorine_levels_en.pdf. Retrieved on 06 Oct 2015

24. IWA Water Wiki (2010) Chlorine residuals contributed by Victoria Beddow. http://www.iwawaterwiki.org/xwiki/bin/view/Articles/ChlorineResiduals. Retrieved on 05 June 2016

25. Department of Human Services (1998) Guidance on the use of rainwater tanks. National Environmental Health Forum, Water series: no. 3, pp 20–21. http://www.ctahr.hawaii.edu/hawaiirain/Library/Guides&Manuals/Guidance_on_the_Use_of_Rainwater_Tanks.pdf. Retrieved on 07 Oct 2015

26. Government of Western Australia (2012) Chlorinated drinking water. Environmental Health Directorate, Department of Health. http://www.public.health.wa.gov.au/cproot/4933/2/chlorinated%20drinking%20water%20finalv2.pdf. Retrieved on 07 Oct 2015

Chapter 8
Rainwater Supply System

Abstract Buildings are infrastructures that can be built in varying shapes and sizes starting from a single-story cubical to multistory high-rise tubular form. Harvested rainwater may be required to be distributed at various locations of a building in conjunction with the distribution system of water from other sources including independent ones. Like normal building water-supply system, a building rainwater-supply system can also be developed according to one of the two common methods: an underground-overhead tank system or a pressurized direct-pumping system. In designing the pumping and piping system, the same hydraulic principles are followed for both a normal and a rainwater-supply system. In this chapter, two methods of creating a water-supply system are discussed, and the design approaches for the elements used in the supply system are described. Finally some safety measures for preventing contamination of water in the supply system are discussed.

8.1 Introduction

In a building, rainwater of varying quality and quantity may be required at various locations where it is needed to be supplied accordingly. A rainwater-supply system is developed in the same way as the normal water-supply system in a building. It is rare to have rainwater as the only source of water for a building; therefore, in most cases a rainwater-supply system is developed separately as a supplemental system alongside of a normal water-supply system. In this chapter, various ways of supplying rainwater and the designing of various components of a rainwater-supply system are described.

8.2 Rainwater-Distribution Approach

A rainwater-distribution system in a building can be developed in two different ways or in a combination of both. The distribution ways are as follows:

© Springer International Publishing Switzerland 2017
S.A. Haq, PEng, *Harvesting Rainwater from Buildings*,
DOI 10.1007/978-3-319-46362-9_8

1. Exclusive rainwater supply and
2. Supplemental water supply.

8.2.1   Exclusive Rainwater Supply

In buildings, rainwater may be needed for some exclusive uses such as drinking or cooking, bathing, toilet flushing, etc. If the availability of sufficient rainwater can be ensured, then rainwater can be supplied for those exclusive purposes without any supplementary supply of water from other sources. In these cases, a single-piping system can be installed as shown in Fig. 8.1 to supply rainwater at various locations where there is demand.

8.2.2   Supplemental Water Supply

In buildings, rainwater may be needed as a supplemental source of water to meet various demands of water consumption. In these cases, a dual-piping system must be installed, as shown in Fig. 8.2, to supply individual fixtures. In this dual-piping system, one pipe will supply rainwater, and the other pipe will supply water from different source other than rainwater. The supply piping for both supply pipes is supported by gate valves.

If in this system one type of water must be protected from contamination by the other type of water, then an extra check valve must be installed on the pipe in which

Fig. 8.1 Single piping for exclusive rainwater supply to a sink

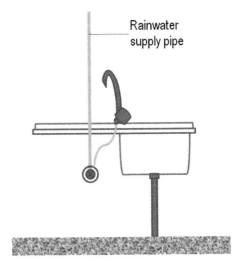

Fig. 8.2 Piping for using
rainwater for flushing

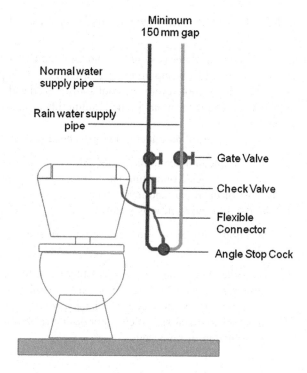

water must be protected. The check valve is placed between the gate valve and the faucet.

8.3 Rainwater-Distribution System

Like a general water-distribution system in a building, rainwater can also be distributed following the same hydraulic principles as in a normal water-supply and - distribution system. The distribution system can be developed according to one of the systems as mentioned below.

1. Underground-overhead tank system and
2. Direct-pumping system.

When rainwater is supplied exclusively, any one of the systems can be adopted. However, when rainwater is supplied as supplemental system along with normal water-supply system, then both of the systems should follow the same distribution principles adopting either an underground-overhead tank system or a direct-pumping system.

8.3.1 Underground-Overhead Tank System

In this system rainwater is stored in underground (UG), the water tank from where the water is lifted by pumping to another tank which is placed on the highest level of the building's roof. From the roof top or overhead (OH) water tank, water is supplied to various locations in the building under gravity through a pipe network as shown in Fig. 8.3.

The advantages of an underground-overhead tank system include the following:

1. It is a simple and economical method.
2. It requires fewest controlling and operating components.
3. Its cost of operation and maintenance is lower.
4. The pressure fluctuation is small.

The disadvantages of an underground-overhead tank system include the following:

1. An overhead tank may not be desirable for aesthetic reason and space scarcity.
2. It imposes an extra load on the building's structural elements.
3. Pressure at the top floors may be inadequate.
4. It involves more tanks, which is not desirable because tanks may be a source of contamination of rainwater.

8.3.2 Direct-Pumping System

In this system, rainwater can be supplied directly from underground or any lower-level reservoir to the points of use, at upper levels in a building, by employing some automatic pressurizing devices in the pipe-network system as shown in Fig. 8.4.

The advantages of a direct-pumping system include the following:

1. It is the most flexible system in terms of available flow and pressure.
2. It requires less floor area.
3. It has low initial cost.
4. It does not impose large weight on the building's structural elements.

The disadvantages of a direct-pumping system include the following:

1. It has the highest operating and maintenance cost.
2. A sophisticated control system and skilled operating staff is needed.
3. The water supply is interrupted due to power failure.
4. The instantaneous pressure fluctuation may be greater.

Fig. 8.3 Underground-
overhead tank
rainwater-supply system

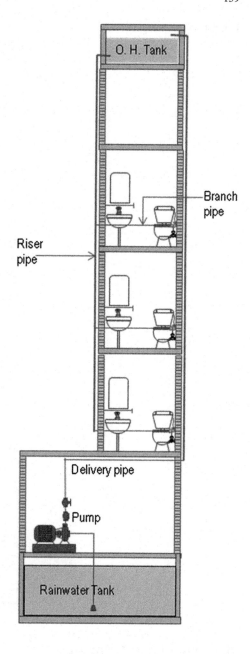

Fig. 8.4 Direct-pumping system

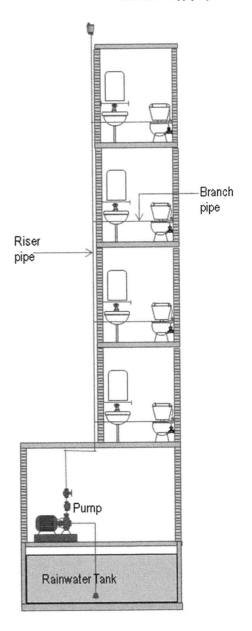

8.4   Pump and Pumping

For a rainwater supply in buildings, a centrifugal pump is the best-suited pressure-developing mechanism. Centrifugal pumps use a circular bowl–shaped volute housing impellers that attaches to and rotates by way of a shaft connected with an electric motor. This pump converts the input power, produced on the shaft created by the motor, to kinetic energy in the water by accelerating it through curved vanes of the impeller. At the start, the pump must be filled with water. Rotation of the impeller creates an outward flow of water due to centrifugal force, which pressurizes water to move outward through the delivery outlet. As water moves outward from the center, a vacuum is created at the impeller center, called the "eye," which causes the suction of water into the pump, which is again moved outward, and the process of outward flow of water is continued through the pump.

A pump must be able to create kinetic energy equal to or a little more than the following energies concerned in the pumping system.

1. Elevation energy
2. Friction energy
3. Pressure energy
4. Velocity energy.

Elevation energy is the energy required to elevate a mass of liquid (rainwater) against the gravitational force gained by a fluid mass due to its position or elevation. Friction energy is energy that is lost to the environment due to the movement of liquid through pipes, fittings, accessories, machineries, etc., in the system. Pressure energy is energy that must be available at the end of delivery pipe, and velocity energy is the combination of liquid mass and the velocity of the moving impellers.

In pumping water, "head" is used to measure the kinetic energy that a pump must create to convey water from one position to another; whereas "flow" is a term used to express the amount of water conveyed per unit time. Both of these factors are mainly used for selecting a centrifugal pump for use in rainwater supply.

8.4.1 Head of Pump

The head of energy, or simply "head" (denoted by "H"), is the energy contained in a water mass per unit weight, which is expressed in meters. It is the energy needed to cause the flow water from one position to other. Therefore, "head" is the measure of kinetic energy that must be created by a pump. The total head also known as "total dynamic head" for a moving pump, which draws water from an underground reservoir to a rooftop overhead tank, is the summation of the suction head, the delivery head, and the friction head loss.

Total dynamic head = total suction head + total delivery head + friction head loss.

When the pump is not running, the total head is termed the "total static head," which is the summation of suction height or static-suction head and delivery height or static-delivery head only as shown in Fig. 8.5.

Suction head: The energy required to lift water from a lower level to the pump-datum level (or the center-point level) of a pump is termed the "suction head"

Fig. 8.5 Total static head of pump

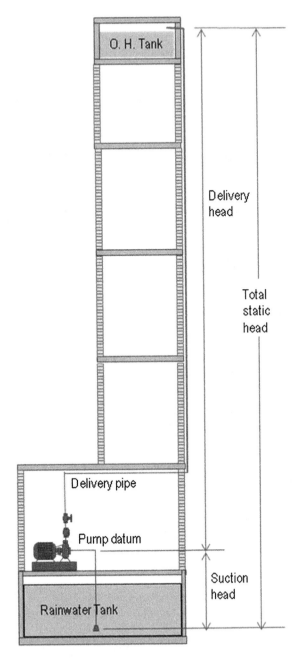

(denoted by "h_s"). To operate a pump effectively, a minimum amount of suction pressure or head is required. This suction head is expressed by the term "net positive suction head" (NPSH).

Total suction head = static suction head + kinetic suction head + suction pressure head.

Delivery head: The energy required in delivering water from the pump datum level to a higher level or further distance is termed the "delivery head" (denoted by "h_d"). The energy contained in the flowing water is the summation of the potential energy, the kinetic energy, and the pressure. Therefore, the total delivery head is denoted by the following formula":

Total delivery head = static delivery head + kinetic delivery head + delivery pressure head.

Static head: Static head is the potential energy contained in water per its unit weight. It is just the vertical height of a water column measured in metres.

Kinetic head: When impellers centrifuge the water, a velocity is imparted to it. This velocity energy is transformed primarily into pressure energy. Therefore, the energy or the head developed is approximately equal to the velocity energy at the periphery of the impeller. This head is termed as the "velocity head" (or the kinetic head) and is expressed by the following formula:

$$H = \frac{V^2}{2g} \tag{8.1}$$

where H is the velocity head developed in metres; V is the velocity at the periphery of the impeller in m/sec; and g is the acceleration due to gravity = 9.80665 or 9.81 m/s^2.

Delivery-pressure head: It is a head caused by any pressure that might be acting on the water in the discharging tank including atmospheric pressure. This is also called the "surface pressure head." When the discharge tank is open to atmospheric pressure, then the delivery pressure head is considered to be zero, i.e., there is no delivery-pressure head.

8.4.2 Flow of Pump

Flow of pump is the quantity of water being flowed per unit time, which is generally expressed in litres per second (lps) or litres per minute (lpm), etc. A centrifugal pump operating at certain speed can provide a varied range of flow depending on the amount of head generated. In general, pumps have an increase in flow when there is a decrease in head.

In case of an underground-overhead tank water-supply system, to determine the flow requirement for pumping it is necessary to know the volume of water to be filled in the reservoir within a particular span of time.

$$\text{Required flow of pump } Q = \frac{V}{T} \tag{8.2}$$

where V is the volume of water to be filled in litres; and T is the time of filling.

Example:
Consider that an overhead tank storing a volume of 12,000 l of water must be filled in 1 h.

Then the flow requirement for pumping = 12,000/60 = 200 lpm.

If that amount of water must be filled in 2 h, then the flow requirement would be 12,000/(2 × 60) = 100 lpm. In this case, the flow requirement is exactly halved by doubling the time for filling the overhead tank.

In the case of a pressurized or up-feed water-supply system, the flow requirement will depend on the number of faucets and water-consuming appliances to be served. In this case, the probable flow rate is considered instead of total flow rate to be required for the desired number of faucets to be served at a time. The probable flow rate of water is the estimated flow of water in a pipe to supply more than one faucet. Generally the probable flow rate decreases with an increase in the number of faucets to be supplied. To determine the probable flow rate of multiple fixtures of varying water supply–fixture units, the Hunter curve is used, which is shown in Fig. 8.6. Water supply–fixture units of various plumbing fixtures are shown in Table 8.1.

8.4.3 Frictional Loss in a Pipe

As water flows through a pipe, there is a resistance due to friction between the flowing water and the inner surface of the pipe, pipe fittings, and other appurtenances installed on the pipe. Friction loss is the loss of energy or head that occurs in a pipe flow due to friction. Friction loss ultimately causes a loss of pressure and results in reduced flow. Friction loss in a pipe is the function of the velocity or rate of flow and the size of the pipe (diameter), the length of the pipe, and the roughness of the inside surface of the pipe. The degree of pipe roughness is determined by the

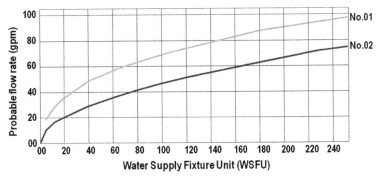

Fig. 8.6 Hunters curve to estimate probable flow rate

Table 8.1 Table for finding pipe size and water supply fixture unit for various plumbing fixtures

Sl. No.	Fixtures		Pipe size (mm)	Fixture unit					
				Private (individual dwelling)			Public (general use)		
				Cold	Hot	Total	Cold	Hot	Total
1.	Ablution tap		12	0.75	0.75	1	1.5	1.5	2
2.	Wash basin		12	0.75	0.75	1	1.5	1.5	2
3.	Shower stall		12	1.5	1.5	2	3	3	4
4.	Bath tubs or combination bath/shower		12	1.5	1.5	2	3	3	4
5.	Sinks	a. Kitchen	12	1.5	1.5	2	–	–	–
		b. Service		–	–	–	2.25	2.25	3
6.	Bidet		-	-	-	3	-	-	-
7.	Water closet	a. Flush tank	12	3	–	3	5	–	5
		b. Flush valve	25	6	–	6	10	–	10
8.	Urinals	a. Flush tank	12	–	–	–	3		3
		b. 20-mm flush valve	20	–	–	–	5		5
9.	Laundry machine	a. 3.6 kg	12	1.5	1.5	2	–	–	–
		b. 7.3 kg		–	–	–	3	3	4
10.	Drinking fountain		9.5	0.25	–	0.25	0.25	–	0.25
11.	Dishwasher			–	1	1	–	–	–

Source [4, 5]

C-factor for the pipe. This is a factor or value used to indicate the roughness of the interior of a pipe. A higher *C*-value indicates higher smoothness and thus lower friction loss. This factor is used as a coefficient in the *Hazen–Williams formula* for determining the flow. For most pipe materials, the *C*-value ranges between 90 and 140. To determine the friction loss in various pipes having various degrees of roughness conditions, different charts are created to determine the friction loss in pipes. A chart for determining friction loss in fairly rough steel pipe is shown in Fig. 8.7.

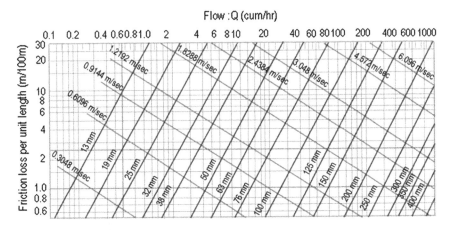

Fig. 8.7 Friction loss in steel pipe. *Source* converted and modified from source [1]

8.4.4   Efficiency of a Pump

No centrifugal pump can run at 100 % efficiency. A general idea is that medium and larger centrifugal pumps offer efficiencies of 75–93 % and those of smaller pumps usually range between the 50 and 70 % efficiency [2]. For centrifugal pumps, the efficiency tends to increase with the flow rate up to a point midway through the operating range (peak efficiency) and then declines as the flow rate increases further. The overall efficiency of a centrifugal pump is simply the ratio of the water (output) power to the shaft (input) power and is illustrated by the equation below.

$$\text{Efficiency of pump } E_p = \frac{P_w}{P_s} \text{ unitless} \tag{8.3}$$

where P_w is the water (output) power in kilowatts; and P_s is the shaft (input) power in kilowatts.

A pump operating below the range of 70 % of its best efficiency point should not be used [3].

8.4.5   Power of a Pump

An electric motor attached to the centrifugal pump consumes energy in one form, i.e., electrical energy, and converts it to a different form resulting in power, which is used to rotate the motor shaft speedily to drive the pump through the coupling with the pump shaft.

"P_w" is the power imparted on water to be lifted or supplied by the pump. It is also termed as "horse power" (HP).

$HP = P_w = w \times Q \times H$ kilowatts (kiloNewton per second) $= 0.0098QH$ (8.4)

where w is the unit weight of the water $= 0.0098$ kN/litre; Q is the flow in lps (litres per second); H is the total dynamic head in metres; and P_s is the power provided to the pump shaft to drive the pump. It is expressed in "brake horse power" (BHP).

$$BHP = P_s = 0.0098QH/E_p \text{ kilowatts} \qquad (8.5)$$

8.5 Pipe Sizing

To develop a rainwater-supply system in a building, five categories of piping for different functional purposes must be designed. Designing of pipes mainly involves determining the pipe size to be used. The different categories of pipes to be designed are as follows:

1. Suction pipe for pump.
2. Delivery pipe to lift water from underground reservoir to overhead tank.
3. Up-feed riser pipe to supply faucets at various locations.
4. Down-feed riser pipe to supply faucets at various locations.
5. Branch piping to supply faucets at particular locations.

8.5.1 Sizing Pipe for Suction and Delivery

The selection of pipe for suction and delivery is based on careful economic analysis considering friction loss in the pipe and the cost of electricity. In the selection of riser and branch pipe, consideration of allowable friction loss in the respective pipe is very effective. For quick selection of pipe size, the easiest way is to consider the velocity of flow in the pipe. The velocity limitation of water flow in metal pipe is 1.2–2.4 m/s. In selecting pipe size for suction and delivery, the velocity of flow should not be considered >1.2 m/s.

Example

Let us find the size (diameter) of a delivery pipe having a flow requirement of 10 lps at a minimum velocity of 1.2 m/s.

$$\text{Flow } Q = AV \qquad (8.6)$$

Therefore, the area of pipe $A = Q/V = \pi \times D^2/4$
Then the diameter of pipe $D = \sqrt{4Q/\pi V}$
The area of pipe needed $= 10/1200 = 0.00833$ m^2 $= 8333.33$ mm^2.

Therefore, the diameter of pipe will be $\sqrt{8333.33 \times 4/\pi} = \sqrt{10{,}610.33} =$ 103 mm.

It must be noted that the size of commercially available pipe is closest to the required pipe diameter of 100 mm. Thus, if 100-mm delivery pipe is used, the flow will be a bit less than 10 lps at a velocity of 1.2 m/sec, as shown in Table 8.1, or the flow will be 10 lps at velocity, a bit higher than 1.2 m/s, depending on pump pressure.

The size of suction pipe should be one size larger than that of the delivery pipe or at least same as the size of the delivery pipe. In this case, a 100 mm–diameter suction pipe is a good selection.

8.5.2 Sizing of Pipe for Riser and Branch

In determining the size of pipe for riser and branch based on the velocity limitation of water flow, it is necessary to know the flow load in the corresponding pipe expressed in terms of number of water supply–fixture units (WSFU) to be served. The water supply–fixture unit for any plumbing fixture is a measure of the probable demand of water by a particular type of plumbing fixture. The value depends on the volume of water supplied, the average duration of a single use of the fixture, and the number of uses per unit time. The minimum size of pipes to supply faucets of various fixtures is more or less fixed. The pipe size and water supply–fixture unit for various fixtures are listed in Table 8.1 and the required pipe size with respect to water supply-fixture units at particular velocity of flow, is shown in Table 8.2.

Table 8.2 Table for determining pipe size based on velocity limitation of water flow for galvanized iron and steel pipe

Nominal size (mm)	For velocity 1.2 m/s				For velocity 2.4 m/s			
	Flow (lps)	Load (wsfu)[a]	Load (wsfu)[b]	Friction (Pa/m)[c]	Flow (lps)	Load (wsfu)[a]	Load (wsfu)[b]	Friction (Pa/m)[c]
12.70	0.23	1.5	–	172.3	0.47	3.7	–	651.5
19.00	0.42	3.0	–	126.1	0.84	8.4	–	472.8
25.40	0.68	6.1	–	96.7	1.36	25.3	7.7	361.5
31.80	1.17	17.5	6.0	71.5	2.34	77.3	23.7	269.0
38.10	1.60	37.0	9.3	60.9	3.20	132.3	52.0	227.0
50.80	2.63	93.0	29.8	46.2	5.27	293.0	171.6	176.5
63.50	3.77	174.0	75.6	37.8	7.54	477.0	361.0	142.9
76.20	5.80	335.0	209.0	29.4	11.60	842.0	806.0	113.5
102.0	10	688.0	615.0	23.1	20.01	1930.0	1930.0	86.2

Source [4]

[a]Applied to piping that does not supply flush valve

[b]Applied to piping that supplies flush valve

[c]Friction loss 'p', corresponding to flow rate, for piping of diameter 'd' having fairly smooth surface condition after extended service from the formula $q = 4.57\ p^{0.546} d^{2.64}$

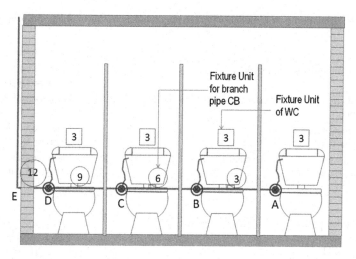

Fig. 8.8 Branch pipe supplying rainwater to water closets in a toilet

Example 1: Let us consider a toilet having four water closets as shown in Fig. 8.8. The branch pipes supplying the fixtures must be sized considering the velocity of flow 1.2 m/s. The sizing approach is listed in Table 8.3.

Designing of pipe for a vertical riser for supplying the branch supply pipes can be performed using the number of water supply–fixture units to be served. In this case, a velocity of flow 2.4 m/sec is suggested to be considered. In a down-feed system, sufficient pressure to satisfy the faucet pressures at the top-level floors is not inadequate due to the limited height of the overhead tanks. In this situation, in the upper one or two floors velocity of the flow in riser pipe can be considered to be 1.2 m/s.

Example 2: Let us consider toilets of a four-story building located in the same vertical alignment. There are four water closets in each toilet as shown in Fig. 8.8. Therefore, the fixture unit of the supply pipe to these toilets is 12 as calculated in Example 1. Now the vertical riser pipe supplying all of these branch pipes must be designed. The sizing approach is listed in Table 8.4.

Table 8.3 Sizing of branch pipes for the toilet shown in Fig. 8.8 using velocity of flow

Branch Pipe	Fixture unit (FU)	Pipe size (mm)
BA supplying one WC	3	12
CB supplying two WCs	3 + 3 = 6	19
DC supplying three WCs	3 + 3 + 3 = 9	25
ED supplying four WCs	3 + 3 + 3 + 3 = 12	25

Table 8.4 Sizing riser pipe for toilets of a four-story building as shown in Fig. 8.9 using velocity of flow

Riser pipe	Fixture unit (FU)	Pipe size (mm)
DE supplying toilet in ground floor	12	25 mm (considering velocity = 2.4 m/s)
CD supplying toilet in first and ground floor	12 + 12 = 24	25 mm (considering velocity = 2.4 m/s)
BC supplying toilet in second, first, and ground floor	12 + 12 + 12 = 36	32 mm (considering velocity = 1.2 m/s)
AB supplying toilet in third, second, first, and ground floor	12 + 12 + 12 + 12 = 48	40 mm (considering velocity = 1.2 m/s)

Fig. 8.9 Riser pipe
supplying toilets in a
four-story building

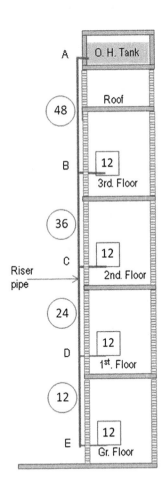

8.6 General Requirement

In the case of a dual-piping system for supplying different qualities of water using harvested rainwater along with water from other sources, some precautionary measures must be taken for safe uses of the water. The following measures are to be strictly followed for this purpose.

1. In duel-piping system where a separate piping is installed for rainwater and water from other sources, the pipe conveying nonpotable rainwater should be labeled and colour-coded to indicate the nonpotable use of rainwater. If rainwater is used for nonpotable purpose, then the pipe should be painted a shade of purple.
2. Cross-connections between the potable water- and nonpotable water-supply system must be avoided.
3. Permanent signage indicating nonpotable rainwater, e.g., "Not for Potable Use" must be provided at every outlet and point of contact. In addition, biodegradable dyes may be added to nonpotable rainwater to alert the user.

References

1. The American Society of Refrigerating Engineers (ASRE) (1989) ASHRAE Handbook, Fundamentals Volume, 1989, cited in http://mte.kmutt.ac.th/elearning/Air_Conditioning/table6.1.htm. Retrieved on 4 Apr 2016
2. Pump and System (2012) 'Centrifugal pump efficiency—what is efficiency?' Contributed by Joe Evans. http://www.pumpsandsystems.com/topics/pumps/pumps/centrifugal-pump-efficiency-what-efficiency. Retrieved on 3 Dec 2015
3. Robert Perez of www.pumpcalcs.com in Pump Fundamentals. http://www.pumpfundamentals.com/centrifugal-pump-tips.htm. Retrieved on 19 Oct 2015
4. Deolalikar SG (1994) Plumbing: design and practice. Tata McGraw-Hill Publishing Company Ltd, Delhi
5. Harris CM (1991) Practical plumbing engineering. Mc. Graw-Hill company

Chapter 9
Groundwater Recharging

Abstract Where there is scarcity of water, it is understood that the dependability on surface water has almost been lost and that the availability of groundwater is either becoming very limited or costly. In such situation where there is potentiality of rainwater harvesting due to having substantial rainfall, there might be sufficient rainwater for using it for some purposes in building projects other than general purpose of uses of rainwater. Improving the groundwater situation using rainwater is considered to be the most effective way of using rainwater other than the use of rainwater for general purposes. Groundwater recharging is the process by which the prevailing groundwater situation is improved in terms of quantity and quality. Rainwater collecting from various catchments of a building can also be effectively used for recharging groundwater through facilitating its percolation into the ground around the building. There are various ways of recharging groundwater using rainwater. Utmost care is taken so that the groundwater is not contaminated through the process of recharging. In this chapter, the soil parameters influencing groundwater recharging and the various ways of recharging are described elaborately. The precautionary measures to be addressed while recharging are pointed out. Finally, the probable impacts of groundwater recharging around buildings by using rainwater, both positive and negative, are also mentioned herein.

9.1 Introduction

Groundwater is one of the major sources of water. In many areas on the globe, groundwater in aquifers is being exhausted continually due to over-pumping and the limiting scope of natural recharging of rainwater percolating into the ground by covering the natural ground surface with hard impervious materials. As a result, groundwater is dwindling, and water levels in aquifers are decreasing. Over the course of time, aquifers are becoming devoid of water. As a remedial measure, rainwater available on roofs and other catchments of buildings can be recharged into those voided aquifers. In this way, the water of the recharged aquifer can be used gainfully at the time of need. There are various ways of recharging

© Springer International Publishing Switzerland 2017
S.A. Haq, PEng, *Harvesting Rainwater from Buildings*,
DOI 10.1007/978-3-319-46362-9_9

groundwater using rainwater. In this chapter, all of those ways of recharging, as well as the measures to be taken in this regard, are presented.

9.2 Groundwater Recharging

Groundwater recharging is the process of augmenting the quantity of the prevailing groundwater and improving its existing quality through facilitating the percolation of water, including rainwater, using by various man-made artificial and natural means. Although groundwater recharging is believed to have considerable positive impacts, unauthorized and uncareful recharging may also pose negative impacts. The anticipated impacts of groundwater recharging by rainwater are discussed below.

9.2.1 Positive Impacts of Recharging

In areas where the withdrawal of groundwater is greater than the rate of recharge, an imbalance in the groundwater reserve is created. The recharging of aquifers is undertaken with a view to achieving the following objectives.

1. To maintain or augment the natural groundwater as an economic resource [1]
2. To facilitate the conservation of excess rainwater underground for subsequent use [1]
3. To prevent the progressive depletion of groundwater levels
4. To combat unfavorable salt balance and saline-water intrusion in the groundwater
5. To improve quality of underground water with good-quality rainwater [1]
6. To improve the yield of an aquifer significantly [1]
7. To decrease and stop land subsidence [1]

Groundwater recharging with rainwater is nothing but storing unused rainwater into the ground during wet periods or periods of low demand and recovering stored groundwater during dry periods, or in times of high demand, to meet both the present and future demand of water either fully or partially. Rainwater recharging is a promising solution that helps to qualitatively improve contaminated groundwater aquifers by decreasing the concentration of pollutants by dilution [2]. Recharging rainwater into brackish aquifers has been found to improve the ambient water quality of the aquifer.

9.2.2   Negative Impacts of Recharging

If groundwater recharge is not carried out properly with the utmost care, there might be negative impacts on the underground and ground environment as well. Depending on the type and quality of injected rainwater and the land geology, the potential for endangering the groundwater may increase. The anticipated negative impacts are as follows.

1. Due to the lack of any financial incentives for building owners to abide by the prevailing laws and regulations regarding the regular routine maintenance of recharge structures, it may cause disrepair or even damage of structures, ultimately turning the aquifer into a source of contaminated groundwater.
2. Mass movements, such as slumps, slides, or earth flows, may occur due to groundwater recharging, thus causing damage to nearby surrounding structures [3].
3. Pathogenic organisms may be introduced into a recharged aquifer if the injected rainwater is not properly disinfected [4].
4. The possibility of formation of disinfection byproducts (DBPs) in situ increases [4].
5. Unless a significant volume of rainwater can be injected into an aquifer, groundwater recharge may not be economically feasible [5].
6. A large space may be required for a system of collection and recharge.
7. Over-recharging of groundwater may cause problems with other underground structures.

9.3   Information Needed

To undertake a groundwater-recharging program at a particular site, the following information should be considered. The decision for installing a recharge structure should be taken based on the analysis of these data, whether already existing or those collected during investigation,

1. Investigate the presence or absence of impermeable layers or lenses that can impede percolation.
2. Understand the surface infiltration rate and hence the recharge rate to suggest potential recharge zones along with the groundwater flow direction within the targeted area.
3. Determine the present depth of the water table as well as the previous groundwater level.
4. Determine the effect of rainwater on the quality of the groundwater aquifer.
5. Estimate the possible volumetric augmentation in the groundwater due to recharge by rainwater.

9.4   Recharge Structure–Design Parameters

The hydraulic design of artificial recharge structures is governed by the following parameters:

1. Infiltration rate
2. Retention time
3. Effective porosity
4. Permeability and
5. Hydraulic transmissivity.

9.4.1 Infiltration Rate

Infiltration is the process of water flow to and through unsaturated soil. Infiltration is characterized as an unsteady state of flow, i.e., the flow rate varies with time under a given head. Therefore, the infiltration rate increases with deeper vertical permeability of the infiltration zone. It should be designed to allow the recharging scheme to operate over a long period at an acceptable rate without excess clogging of the infiltration zone. The design rate thus should be less than the maximum infiltration rate. The infiltration rate of any soil is determined testing that soil for infiltration. In the absence of infiltration rate data, Table 9.1 can be used to make assumptions.

Table 9.1 Infiltration rates for different type soils

Soil type	Infiltration rate(cm/h)
Gravelly silt loam	12.60
Clay loam	10.11
Silt loam	5.31
Sandy loam	4.90
Clay (eroded)	4.52
Sandy clay loam	3.61
Silty clay loam	1.83
Stony silt loam	1.40
Fine sandy loam	1.40
Very fine sandy loam	1.29
Loam	1.27
Sandy clay	0.13
Heavy clay	0.05
Light clay	0.00
Clayey silt loam	0.00

Source Free et al. (1940), Cited in [6]

The quality of infiltrating rain or stormwater also plays vital role in infiltration rate. Clear rainwater will infiltrate at a comparatively faster rate and for a longer period of time than contaminated rainwater. Therefore, infiltrating rainwater should be largely free from suspended and colloidal matter. Stormwater may need to be pretreated to remove excess turbidity.

9.4.2   Retention Time

To ensure hygienic safety of the recovered water, artificial-recharge schemes should be designed to provide a retention time of at least 3 weeks and preferably 2 months [7]. The retention time is controlled mainly by the infiltration rate and the hydraulic transmissivity of the formation.

9.4.3   Effective Porosity

Effective porosity, also known as "kinematic porosity," is that portion of the total void space of a porous soil material that is capable of transmitting water. Total porosity is defined as the ratio of the total void volume to the total bulk volume. Traditionally, porosity ratios are multiplied by 100 and expressed as a percent. There might be pores filled with water adsorbed on clay particles or other grains, particularly in saturated porous soil, through which water does not flow. Total porosity is the total void space in the soil through which water flows or does not flow. Unconnected pores are often termed as "dead-end pores". Some water is also contained in interconnected pores, which is held in place by molecular and surface-tension forces. This "immobile" water volume also does not participate in water flow. Therefore, "effective porosity" denoted the void space formed by the interconnected voids through which water flows. Effective porosity is always less then total porosity. The effective porosity can be determined by using the formula [8] given below.

$$P_e = \frac{V_{mw}}{V_1} = \frac{V_{mw}}{V_s + V_{mw} + V_{iw}} \tag{9.1}$$

where

V_{mw} is the 3D volume of the mobile pores containing water;
V_{iw} is the 3D volume of pores containing immobile water adsorbed onto the soil particle surfaces and the water in the dead-end pores; and
V_s is the 3D volume of the solid phase.

Table 9.2 Permeability of
some types of soil and rock
(Campbell and Lehr 1973)

Type of rock	Permeability (m/d)
Fine sand	1–5
Coarse sand	20–100
Gravel	100–1000
Mixed sand and gravel	50–100
Sandstone	0.1–1.0
Clay	0.01–0.05
Shale	Negligible
Limestone	Negligible
Fractured or weathered rock	0–30
Solid rock	Negligible

Source Campbell and Lehr (1973), cited in [7]

9.4.4 Permeability

The permeability of a formation is the flow rate at which water moves through the
voids under a head of 1 m across a distance of 1 m. The main factors affecting
permeability are the effective porosity and the degree at which the pores in the
formation are interconnected. Table 9.2 lists the permeability of some types of rock.
Permeability is usually expressed either as a permeability rate in mm/h (millimetres
per hour) or m/d (metres per day) or as a coefficient of permeability, denoted by
"k", in m/s (metres per second).

9.4.5 Hydraulic Transmissivity

The hydraulic transmissivity is the product of permeability and the water-filled
depth of the formation, and it denotes the water-transmission capacity of the soil
formation. With permeability k and formation depth D, hydraulic transmissivity is
often given as the kD factor, which is expressed in cum/day/m width (cubic metres
per day per width in metres) [7].

9.5 Methodology of Recharging

Groundwater in aquifers can be recharged in two different ways:

1. Natural recharging and
2. Artificial recharging.

9.5.1 Natural Recharging

In natural recharge, the rainwater or surface water percolates into a shallow and deep aquifer by itself through soil pores of uncovered soil surfaces and fissures of rock mass. To help developing groundwater recharging naturally, various simple techniques can be adopted as mentioned below.

1. Keeping the ground surface green
2. Rainwater spreading on ground
3. Constructing water bodies
4. Facilitating rainwater to percolate into the ground
5. Induced recharge from storm-drain appurtenances.

Keeping the ground surface green: The most fundamental and simplest way of promoting the natural recharge of groundwater is to keep the ground surface around the building green as much as possible. This will allow direct percolation of rainwater into the soil and ultimately will recharge the groundwater. Unnecessarily creating hard surfaces should be avoided. An internal metal road should be made as narrow as possible.

Rainwater spreading on surface: Discharging excess rainwater on the ground surface is generally proposed where a green surface is found available. To avoid erosion of the soil surface, a confined stone bed under rainwater-down pipes is proposed as shown in Fig. 9.1. In unavoidable circumstances, various others measures are suggested to facilitate seepage of rainwater into the ground while draining rainwater through sewers.

Constructing water bodies: Where possible, water bodies in the form of pond or lake, whether small or large, should be incorporated in the landscape around a building or in a building complex. Depending to soil conditions, if water can be retained in any water bodies for a long period, then its sides and bottom should not have any impermeable lining. These water bodies will help storing water, i.e., rainwater and at the same time recharge the groundwater, albeit slowly.

Facilitating rainwater seepage into the ground: In unavoidable cases where metal surface is to be built around a building, additional measures taken in the metal surface can facilitate the percolation of rainwater into the ground. Metal surfaces such as parking lots, open yards, etc., can be built with pre-cast slabs having holes, as shown in Fig. 9.2, and keeping sufficient gap around solid slabs, both of which will facilitate percolation of rainwater into the ground as well as prevent accumulation of stormwater on the surface.

Induced recharge from storm-drain appurtenances: While disposing of excess rainwater from a building or stormwater in a building complex, through drainage piping the rain or stormwater can be facilitated to percolate into the ground by taking some special measures as described below. These measures may not be considered as the recharging methodology, but they obviously will facilitate the percolation of whatever quantity of rainwater into the ground.

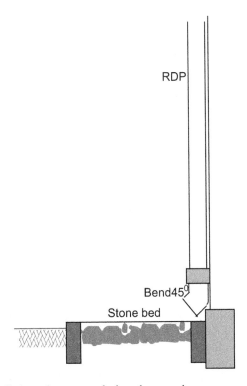

Fig. 9.1 Rainwater discharged on a stone bed on the ground

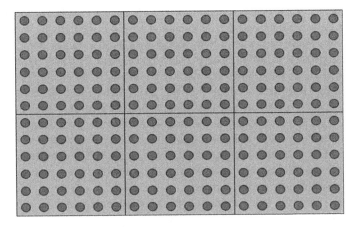

Fig. 9.2 Metal surface built of pre-cast slab with holes

Recharging through drain pipes: When draining out rainwater through any underground sewer, it is suggested to use special pipe in which the bottom-half perimeter should remain cut at approximately 150 mm intervals as shown in

Fig. 9.3 Half perimeter–cut
rainwater-drainage pipe

Fig. 9.3. This is done to facilitate rainwater to percolate into the ground while flowing through the pipe. Conventionally perforated pipes have been suggested in literature, and these are generally factory made. However, the proposed cut system can be done on-site. Furthermore, cutting the half-perimeter of pipe will produce more open area than perforating the same area of pipe.

Recharging through an inspection pit: In the inspection pits of any drainage system, the bottom slab is generally a solid element made of reinforced cement concrete. To facilitate the intrusion of rainwater, brick chips or gravel, as shown in Fig. 9.4, can be placed instead of putting a solid slab underneath. This type of inspection pit can be termed a "recharge inspection pit."

9.5.2   Artificial Recharging

Artificial recharging of an aquifer is the process of planned infiltration of rainwater or surface stormwater into the aquifer by constructing simple civil structures. This concept of channeling surface rainwater into the ground aquifer can be accomplished by the following various ways independently or making a combination of two or more ways. There are various ways of artificial recharging; however, not all of them are suitable for performing on inside-building premises or a complex. Following are the popular techniques of artificial recharging those can be constructed around the buildings collecting rainwater from the catchments on or around the buildings.

1. Recharge pit
2. Recharge trench
3. Recharge troughs
4. Recharge well and
5. Injection well.

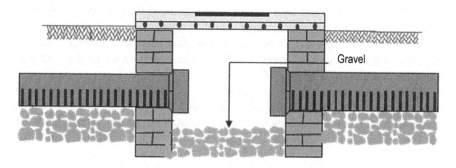

Fig. 9.4 Gravel at the bottom of a stormwater-inspection pit

Before considering the adoption of any type of artificial recharging following factors should be considered for its sustainability [9].

1. Availability of sufficient quantity of rainwater
2. Quality of stormwater or rainwater available
3. Resulting water quality after reactions with existing water and aquifer materials
4. Clogging potential of soil
5. Underground-storage volume available
6. Depth to underground storage space
7. Transmission characteristics
8. Topography and applicable methods (injection or infiltration)
9. Legal and institutional constraints
10. Cost of construction, operation, monitoring, and maintenance
11. Cultural and social considerations.

Recharge pit: In alluvial areas where there is permeable soil either remaining exposed on the land surface or located at very shallow depth, rainwater recharging can be performed using a small structure called a "recharge pit." These pits are constructed generally for recharging shallow aquifers. This technique is suitable for the amount of rainwater that can be collected from a building having catchments of $100 \, m^2$ [10].

Recharge pits may be of any shape: square, rectangular, or round. They are generally constructed $1-2 \, m^2$ or round and approximately $2-3 \, m$ deep. The pits are filled with boulders of size $50-200 \, mm$, gravels of size $5-10 \, mm$, and coarse sand of $1.5-2 \, mm$ in graded form as shown in Fig. 9.5. Boulders are placed at the bottom of pit; a layer of gravel is placed in between; and a coarse sand layer is placed at the top. The silt content and suspended matters that will come with runoff water remain deposited on the top of the coarse sand layer, which can be easily removed. The thickness of these layers varies from $0.30-0.50 \, m$ depending on the silt load in the rainwater. For a smaller roof area, a pit may be filled with broken bricks or cobblestones.

On a roof, mesh on the gutters and a dome-shaped grating on the inlet of rainwater pipe should be provided so that leaves or any other solid waste or debris is prevented from entering the pit. A de-silting or collecting chamber may also be provided at the ground to arrest the flow of finer particles to the recharge pit. The top $50-100 \, mm$ of the sand layer should be scrapped and then be replaced by fresh sand to maintain the constant recharge rate through the filter material. First-flush arrangement should be provided before the collection chamber to drain off the first showers. Perforated cement Concrete tile should be placed on top over the sand to avoid erosion as shown in Fig. 9.6.

Designing a recharge pit: The size of recharge pits to be constructed depends on the catchment area, rainfall, infiltration rate, etc.; the depth of the pits can be estimated by the ratio of maximum amount of rainfall on the site to the area of the pit. For a round pit, the diameter is calculated using the following equation [11].

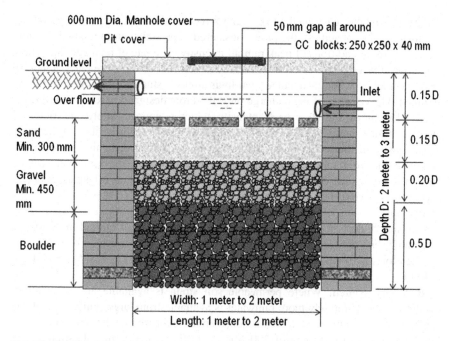

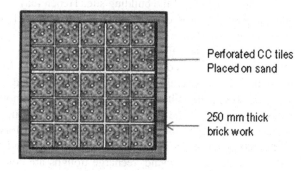

Fig. 9.5 Details of a recharge pit

Fig. 9.6 Perforated concrete
tiles on sand in a recharge pit
(plan)

$$h = \frac{1}{a}(A \times C \times R - K \times a) \qquad (9.2)$$

where

a is the area of the pit (fixed);
h is the depth of the pit;
A is the catchment area of the rainwater;
C is the runoff coefficient of the catchment;
R is the average annual rainfall; and
K is the infiltration or recharge rate.

There is another easy method of sizing a rectangular or square recharge pit. In this method, the capacity of the pit is considered to be sufficient to retain the runoff occurring from conditions of peak rainfall intensity. The rate of recharge in relation to the runoff is a critical factor. However, because accurate recharge rates cannot be made available without detailed geo-hydrological studies, the rates must be assumed. The capacity of the recharge pit is therefore designed to retain runoff from at least 15 min of rainfall at peak intensity, which is approximately one fourth of the peak hourly rainfall [12]. The pit is filled with porous material, so a factor of loose density of the media (void ratio) must be applied in designing the pit size. Although the void ratio of the filler material varies with the kind and size of material used, in case of using very common materials—such as brickbats, pebbles, and gravel—a void ratio of 0.5 may be assumed.

Example

Consider a catchment area of 200 m^2, the peak intensity of rainfall in 15 min is 100 mm/h, and the runoff coefficient of the catchment considered to be 0.85.

Therefore, a volume of recharge pit required = 200 $\times$ 0.1 $\times$ 0.85/0.5 = 3.4 m^3 (3400 L)

Recharge trench: Where permeable soil strata is available at very shallow depths and comparatively more rainwater is available from larger building roofs or catchments, then a recharge trench is considered. Recharge trenches are suitable for buildings having a roof area of 200–300 m^2 [10] and the permeable strata occur within a depth ranging between 2–5 m. Generally recharge trenches are constructed around the periphery of the building site. Trench may be constructed 0.5–1 m wide and maximum 5 m deep [13]. The length of trench will be depending on the availability of rainwater to be recharged.

The trench is backfilled by boulders of size 50–200 mm, gravel of size 5–10 mm, and coarse sand of size 1.5–2 mm in graded form. Boulders are placed at the bottom layer, gravel in between, and coarse sand at the top, as shown in the Fig. 9.7, so that the silt content that comes with runoff will be separated on the coarse-sand layer at the top, which can easily be removed. The top layer of sand should be cleaned periodically to maintain a satisfactory recharge rate.

Recharge troughs: To recharge runoff from paved or unpaved areas draining out of a building compound, recharge troughs are constructed to trap the surface runoff. Recharge troughs are generally located at the entrance of a building compound. These structures are similar to recharge trenches except that the excavated portion is not filled with filter materials. These troughs are made 0.5–0.75 m wide and approximately 1–1.5 m deep. Each trough is provided with a number of recharging bores of 100 mm diameter with the number depending on the amount of water to be recharged and the soil permeability. To facilitate speedy recharge, boreholes are drilled at regular intervals of approximately 1500 mm in the trench as shown in Fig. 9.8. In terms of design, there is no need to incorporate the influence of filter materials. Very limited amount of runoff from a ground or paved surface can be recharged through this type of structure for its limitation with regard to size [14].

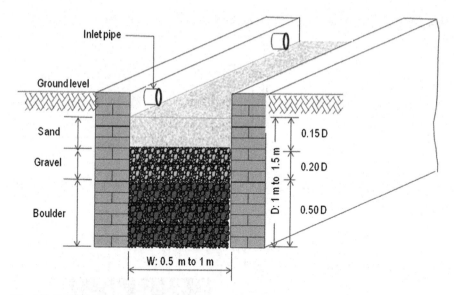

Fig. 9.7 Recharge trench

Recharge wells: Soak wells are the traditional method of disposal of surface water from buildings and paved areas where no drainage exists in the area. The purpose of making these soak wells is to facilitate the infiltration of surface water into the ground. This concept is partially applied in recharge wells to recharge groundwater by rainwater and stormwater. In a recharge well, additional care is taken where filtered stormwater is allowed for recharging through it; however, in soak wells some form of filter media is put into the well to separate or absorb the impurities present in the stormwater before infiltration into the surrounding soil.

A recharge well is preferred where the aquifer is located at comparatively shallow depth under clayey strata. A recharge well, as shown in Fig. 9.9, should be designed and constructed in accordance with the surrounding soil conditions where it is to be constructed. The suitability of the soil strata should must be tested according to the percolation-test method.

The size and number of recharge wells to be used for recharging groundwater depends on the following factors.

1. Infiltration rate: This is the rate at which the water can be absorbed into the surrounding soil around the well.
2. Maximum rainfall: The storage volume inside the recharge well should be able to contain the flow from the catchments resulting from maximum rainfall. The volume must be less than the amount of rainwater that can be infiltrated into the soil.

Fig. 9.8 Recharge trough

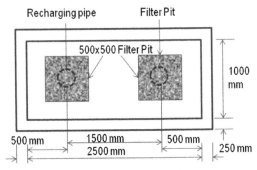

(a) Plan of recharge trough

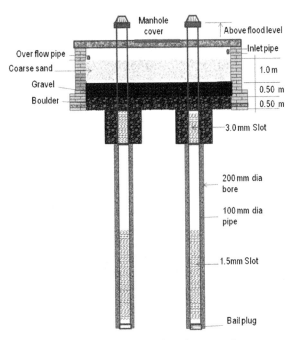

(b) Sectional elevation of a recharge trough

3. Groundwater level: To avoid any contamination of groundwater, the bottom of recharge well must be above the highest groundwater level.

The diameter of recharge or soak well may vary from 0.5–3 m depending on the availability of rainwater to be recharged [15]. In designing soak wells for groundwater recharging by rainwater, the general governing factor is the permeability of the soil. The size of a recharge well can be determined from the following formula [16] assuming that the base is not made solid.

Fig. 9.9 Schematic diagram
of a recharge well

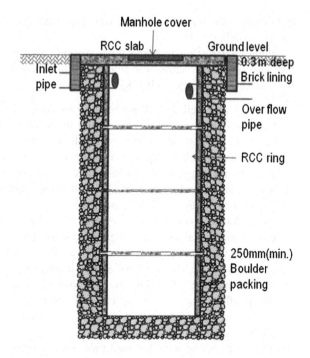

$$q = k(\pi \times r^2 + 0.5 \times \pi \times D \times l) \qquad (9.3)$$

where

q is the dissipation rate of rainwater from the recharge well into the soil in cum/s;
k is the coefficient of the infiltration in m/s;
D is the diameter of the recharge well in m;
r is the radius of the recharge well in m; and
l is the depth of the recharge well in m.

Recharge wells are constructed by digging the earth to a depth depending on the
depth of the well plus the depth of the top of the well below ground. The bottom of
the well should reach the permeable sand strata. Wells may be constructed by
placing precast perforated concrete or clay rings, one above another, or by brick
masonry work. Packing is sometimes provided in the outer space of the well using
shingles or gravels. If normal concrete rings having no perforations are used, then
small openings are kept in the joints between the rings and the brick work to pass
water through these openings.

In determining the depth of the recharge well, the position of the permanent
groundwater table must be considered. It is preferred to keep the bottom of the well
high above the groundwater table. In critical conditions, the bottom of the recharge
well should be at least 0.5 m above the groundwater level [17]. The groundwater

level considered should be the permanent groundwater level up to which the level can be raised at most.

While constructing recharge wells, various precautionary measures should be taken for safe construction. The following safety precautions should be taken while constructing a recharge well [18].

1. The recharge well should not be close to any of walls, foundations, or column footings. The distance of the centre of the recharge well from the footing edge should be at least twice the depth of the footing.
2. The recharge well should not be close to a basement wall or any other underground structure. This may cause water to seep into the basement through the walls during the rainy season. The centre of the recharge well should be at least three times the depth of the basement.
3. The soil should not be made too loose while digging the recharge well. Extreme care should be taken while digging loose soil for construction of deep recharge well. Sufficient safety provisions, such as shoring for excavation and safety belts and helmets, etc., for workers, should be assured.
4. The recharge well should be properly covered while construction is taking place.
5. An overflow pipe must be provided on the recharge well. Extend the overflow pipe up to the nearby surface drain or water body to avoid water from logging around the recharge well.
6. The depth of the recharge well must be restricted above the recorded highest level of the fluctuating water table.
7. The recharge well should not be built if the water table is within 1.50 m from the ground level.
8. The gravel packing should be performed by skilled workers to avoid the recharge well from caving in or collapsing.
9. Sufficient precautionary measures should be taken to ensure that water will not enter into the recharge well from outsides because this may lead to collapse of the well.

Abandoned tube wells: If any tube well is found abandoned due to drying up of the aquifers and a decrease of the water table down to a deeper aquifer, rainwater harvesting through that abandoned tube well can be adopted to recharge the dried-up aquifer.

After diverting the first-flush rainwater from the catchments, the subsequent rainwater should be passed through a filter medium before it is allowed to enter the tube well to be used for recharging as shown in Fig. 9.10. The filtering unit may be like a recharge pit having a bottom sealed by concrete bed. A sump is created at the center of the pit, in which a pipe is connected to convey filtered rainwater to the well.

If the roof area is very large, a sedimentation tank should be employed before the filtering unit. The sedimentation tank is connected to the filter pit by a pipe having diameter equal to the diameter of the outlet pipe from the filter unit. The slope of the connecting pipe should be 1:15. The filter pit can be designed in varying shapes and

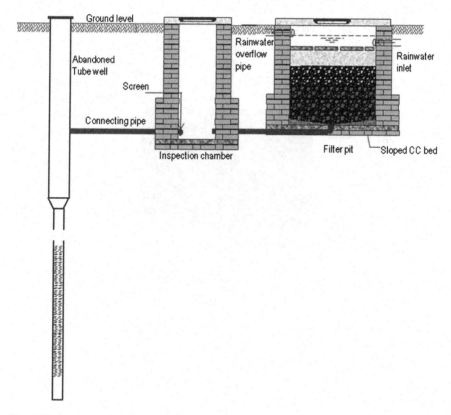

Fig. 9.10 Ground recharging using an abandoned tube well

sizes depending on the available runoff or recharge rate. The filter unit is filled with graded aggregates; a boulder layer placed at the bottom, gravel layer in the middle, and sand layer is at the top with varying thickness ranging 0.30–0.50 m. Another inspection pit should be constructed in between the filtering pit and the tube well to monitor the quality of the filtered water. The inlet of the connecting pipe between the recharge well and the inspection pit should have a screen to prevent any rodents or other unwanted objects from entering the well.

Injection well: Injection wells are preferred for certain areas where natural percolation would be impractical due to the existence of impermeable strata in between the surface and the aquifer. Injection wells are also preferable when available land is scarce where natural infiltration of rain or stormwater is almost impractical [19]. An injection well is the only method for artificial recharging of deep-seated aquifers confined by a very poor permeable or impermeable overburden. Rainwater to be injected in injection recharge well should be disinfected and filtered. Injection wells may be operated under pressure or under gravity.

In nonpressurized injection, well water is allowed to percolate under gravity by passing through a filter bed, which comprises of sand and gravel. A modified

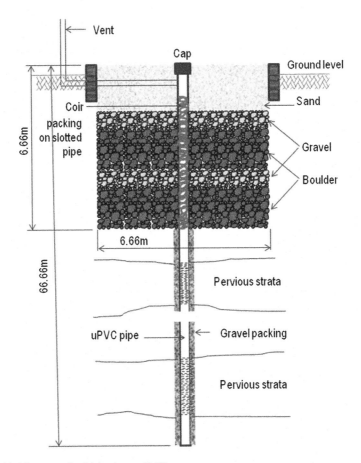

Fig. 9.11 Nonpressurized injection well [1]

injection well is developed by constructing a borehole of 500-mm diameter, which is drilled to the desired depth, preferably 2–3 m below the recorded level of water table in the area. Inside this hole, a slotted casing pipe of 200 mm diameter is inserted as shown in Fig. 9.11. The annular space between the borehole and the pipe should be filled with gravel. The gravel pack must be developed adopting the compressor method until it generates clear water. To stop the suspended solids from entering the recharge tube well, a filter mechanism is provided at the top [13].

In a pressurized injection well, treated and disinfected rainwater is injected under pressure into a confined aquifer as shown in Fig. 9.12. The injection-well pipe can be filled with sand for filtering disinfected rainwater. An additional pump is employed to pressurize the rainwater for passing it through filter media speedily and finally to inject it forcefully into the ground. To restore the function of an injection recharge well due to clogging, a reverse flow can be created by employing vacuum

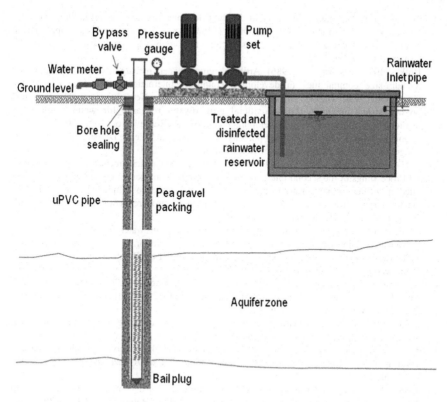

Fig. 9.12 Artificial recharge through a pressurized injection well

pressure, thus eliminating the clogging and improving the efficiency of the groundwater recharge. The main disadvantage of this system is that it requires extra energy, and thereby cost, for operation and maintenance.

References

1. Todd DK (1980) Groundwater hydraulogy, 2nd edn. John Willey and Sons, New York
2. Wateraid in Nepal (2011) Rainwater harvesting for recharging shallow groundwater. p 39 file:///C:/Users/User/Downloads/Rainwater%20harvesting%20for%20recharging%20shallow %20groundwater%20(1).pdf. Retrieved on 19 Oct 2015
3. United State Department of Agriculture (USDA) (1967) Ground-water recharge. Soil Conservation Service, Engineering Division, Technical release no 36. p 17 http://ponce.sdsu. edu/usda_scs_tr36.pdf. Retrieved on 28 Jan 2016
4. Environmental Protection Agency USA, (EPA) (2016) Aquifer Recharge and Aquifer Storage and Recovery http://water.epa.gov/type/groundwater/uic/aquiferrecharge.cfm. Retrieved on 28 Jan 2016
5. Organization of American States (OAS) Artificial recharge of aquifers https://www.oas.org/ dsd/publications/Unit/oea59e/ch18.htm. Retrieved on 28 Jan 2016

6. Free et al (1940) Cited in A Field Method for measurement of infiltration. General Ground-Water Techniques' by A. I. Johnson (1963), Geological Survey Water-Supply Paper 1544-F, U.S. Geological Survey, Washington. http://pubs.usgs.gov/wsp/1544f/report.pdf. Retrieved on 28 Jan 2016

7. Hofkes EH, Visscher JT (1986) Artificial groundwater recharge for water supply of medium-size communities in developing countries. International Reference Centre for Community Water Supply and Sanitation The Hague, The Netherlands. pp 11–12, http://www.samsamwater.com/library/Artificial_groundwater_recharge_for_water_supply_of_medium-size_communities_in_developing_countries.pdf. Retrieved on 3 Dec 2015

8. Argonne National Laboratory Total porosity. Environmental Science Division http://web.ead.anl.gov/resrad/datacoll/porosity.htm. Retrieved on 19 Oct 2015

9. O'Hare MP, Fairchild DM, Hajali PA, Canter LW (1986) Artificial recharge of groundwater. In: Proceedings of the second international symposium on artificial recharge of groundwater, cited in 'artificial ground water recharge with a special reference to India by Amartya Kumar Bhattacharya, IJRRAS 4(2) August 2010. http://www.arpapress.com/volumes/vol4issue2/ijrras_4_2_12.pdf. Retrieved on 19 Oct 2015

10. Ministry of Railways, Government of India (2004) Rain water harvesting' Centre for Advanced Maintenance Technology, Maharajpur, Gwalior, p 30, http://www.rdso.indianrailways.gov.in/works/uploads/File/Handbook%20on%20Rain%20water%20harvesting(1).pdf. Retrieved on 28 Jan 2016

11. Wateraid in Nepal (2011) Rainwater harvesting for recharging shallow groundwater p 14, file:///C:/Users/User/Downloads/Rainwater%20harvesting%20for%20recharging%20shallow%20groundwater%20(1).pdf. Retrieved on 19 Oct 2015

12. Center for Science and Environment (CSE) (2003) A water harvesting manual for Urban Areas Case Studies from Delhi, New Delhi, p 13

13. Bouwer H (2002) Artificial recharge of groundwater: hydrogeology and engineering. Hydrogeology J Doi: 10.1007/s10040-001-018 file:///C:/Users/User/Desktop/eqc09112006_ex18.pdf. Retrieved on 16 Jun 2016

14. Center for Science and Environment (CSE) Components of a rainwater harvesting system http://www.rainwaterharvesting.org/Urban/Components.htm. Retrieved on 19 Oct 2015

15. Central Public Works Department (CPWD) (2002) Rainwater harvesting and conservation, manual. p 24, New Delhi, http://cpwd.gov.in/Publication/rain_wh.PDF. Retrieved on 15 Jun 2016

16. Plastube (Undated) Soakwell information http://www.plastube.com.au/files/Plastube_Soakwell_Brochure.pdf. Retrieved on 20 Jun 2016

17. The National Academy Press (1994) An introduction to artificial recharge. National Research Council. Ground Water Recharge Using Waters of Impaired Quality. p 23, Washington, DC, http://www.nap.edu/read/4780/chapter/3#17. Retrieved on 15 Jun 2016

18. CommonFloor.com (2011) Precautions to be taken while constructing a recharge well. Contributed by Kiram Rao, https://www.commonfloor.com/guide/precautions-to-be-taken-while-constructing-a-recharge-well-6281.html. Retrieved on 19 Oct 2015

19. Environmental Protection Agency (EPA) (2012) Aquifer Recharge (AR) and Aquifer Storage & Recovery (ASR), USA http://water.epa.gov/type/groundwater/uic/aquiferrecharge.cfm. Retrieved on 19 Oct 2015

Chapter 10
Rainwater Drainage

Abstract It is rare that the available rainwater can be harvested fully either for all purpose of use or for groundwater recharging. Therefore, in the majority of cases there would be need for draining of excess rainwater after needful harvesting. Furthermore, provision for rainwater drainage should be incorporated in all types of rainwater-harvesting systems either for overflowing or disposing of rainwater during certain maintenance or repair works. There are various ways of draining rainwater or stormwater. Among the drainage methodologies, a gravity-drainage system is mostly planned around a building or in a building complex. Therefore, gravity drainage by surface and underground piped drains is discussed herein. Among all of the aspects of drainage, the design of drainage system is emphasized in this chapter. Furthermore, the appurtenances of drainage piping are also delineated.

10.1 Introduction

Sometimes a situation may so arise that due to excessive rainfall after a sufficient quantity is collected for storing and recharging, excess rainwater may be needed to be drained out properly. In addition, an overflow pipe should be installed in almost all groundwater-recharge structures, which also must be properly developed including a system of draining out excess rainwater. Furthermore, to avoid the "ponding" of rainwater on the ground surface after a heavy rainfall event, an appropriate drainage system is required. The ways and means of disposing of rainwater is different for different situation and condition. In this chapter, various ways of draining excess rainwater from buildings and its premises, including the drainage of rainwater-harvesting structures, are elaborated.

© Springer International Publishing Switzerland 2017
S.A. Haq, PEng, *Harvesting Rainwater from Buildings*,
DOI 10.1007/978-3-319-46362-9_10

10.2   Draining System

Broadly there are two basic systems for draining rainwater or stormwater. Depending on the site conditions around the building, an appropriate system of drainage is chosen. The complete drainage system may be combination of both systems. The various rainwater-draining systems are as follows.

Considering the position of the disposal point, the drainage system may be typed as follows:

1. Gravitational system and
2. Pumping system.

10.2.1   Gravitational System

Drainage systems are mostly developed with a gravitational system that works under gravitational force; as such, it is an economical system. In flat terrains where the point or level of disposal is lower than the point or level of generation or accumulation of storm or rainwater, then this type of system is applicable. In this system, only the conveying media and, if necessary, various appurtenances are needed to drain rainwater. The conveying system may be of two different types, such as the following:

1. Surface drainage (open or covered system) and
2. Subsurface drainage (closed system).

An open-surface drainage system comprises natural or man-made swales or ditches and is generally used on sites with predominantly natural surfaces in contrast to a closed-subsurface drainage system, which comprises inlets, catch basins, pipes, manholes, and outlets mostly placed underground. In this chapter, mostly closed-drainage system and constructed-surface drains will be covered.

10.2.2   Pumping System

In a situation when the point or level of generation or accumulation of rainwater or stormwater remains lower than the point or level of disposal, then either the lifting or throwing of rainwater or stormwater is required for draining. In this case, some form of energy is necessary to dispose of, or drain out, rainwater or stormwater. Generally a pump is used for these purposes.

Depending on the quality of rainwater or stormwater, a pump of different characteristics may be required. For rainwater disposal, a normal centrifugal pump can be used; however, for stormwater containing heavy suspended particles or matters might require the use of a sewage handling–type pump.

10.3   Gravitational-Drainage Design Factors

The initial job in designing a rainwater-drainage system should estimate the quantity of rainwater to be drained out. The estimation is based on the following factors:

1. Imperviousness of the surface
2. Ground slope and time of concentration
3. Intensity of the rainfall for a design period and
4. Duration of the rainfall.

10.3.1   Imperviousness of the Surface

The percentage of imperviousness of the drainage area may be obtained from available data for a particular area. According to the Central Public Health and Environment Engineering Organization (CPHEEO) manual, in the absence of such data, the values listed in Table 10.1 may serve as a guide.

10.3.2   Time of Concentration

Time of concentration is the time required for stormwater runoff to concentrate and be conveyed from the hydraulically most remote point of a sub-watershed to the point under consideration or an outlet of drainage. Time of concentration is not a constant but varies with depth of flow, slope of surface, and hydraulic condition of the sub-watershed. Time of concentration is the sum of overland flow and the drain flow times from the most remote inlet to the point of design in the drainage system, i.e.,

$$\text{Time of concentration } T_c = T_o + T_d \text{ in minute} \qquad (10.1)$$

where

T_o is the time (in minute) required in overland flow to reach an inlet; and
T_d is the time (in minute) required to flow from the most remote inlet to the point of final disposal,

Table 10.1 Imperviousness factor for different type of areas

Type of area	Imperviousness (%)
Commercial and industrial areas	70–90
Residential areas (high density)	60–75
Residential areas (low density)	35–60
Parks and undeveloped areas	10–20

Source [1]

$$= \sum_{i=1}^{n} \frac{L_i}{V_i}$$

where

L_i is length (in m) of ith number sewer along the longest drainage route; and
V_i is the velocity (in m/min) of flow in ith number of the sewer.

The overland flow time (T_o) varies from 5 to 15 min depending on the overland travel distance, land topography, and land characteristics [2]. The drain flow time (T_d) is calculated from the hydraulic properties of the drainage channel. The flow time must be assessed along several possible routs of drainage network, and the longest drainage route should be identified to find the maximum drain flow time.

At the final disposal point of a drain, the peak runoff occurs when steady rainfall of uniform intensity is received in all parts of the catchment and contributes totally to the outflow at that point. This condition is met when the duration of rainfall becomes equal to the time of concentration [1].

10.3.3   Hourly Intensity of Rainfall

The detail of intensity of rainfall was discussed in Chap. 2. In designing a drainage system for rainwater or stormwater, maximum rainfall intensity data for the locality of the building should be studied to determine the required design parameters. The annual, monthly, or even daily rainfall intensity data are not very effective for the calculation of runoff quantities for vertical or horizontal conveying media. The maximum intensity of rainfall in 1 h is the most important data for the drainage-system designers. When rain falls over a longer period, it is recorded separately for every hour. Peak intensity per hour over the total falling period is then calculated for design purposes. The hourly intensity of rainfall is very much dependent on local conditions. Therefore, for designing drainage structures, the local hourly rainfall intensity must be known.

The intensity of rainfall for storm-water drainage must be calculated at the time of concentration. For a rain event of return period (T) years, the rainfall intensity (I) is the average rate of rainfall from such a rain event having a duration equal to the time of concentration (T_c). The rainfall intensity (I) can be found from the intensity–duration–frequency (IDF) curves by estimating the duration of rainfall matching with the time of concentration T_c, and selecting the required return period (T) in number of years [2]. The hourly rainfall intensity for different part of Bangladesh is listed in Table 10.2.

Table 10.2 Hourly maximum rainfall intensity in some districts of Bangladesh

Serial no.	District	Return period (y)	Hourly maximum (mm/h)
1	Barisal	10	62.71
		25	74.21
		50	82.74
2	Bogra	10	74.59
		25	90.93
		50	103.05
3	Chitagong	10	94.50
		25	113.65
		50	127.86
4	Cox's Bazar	10	80.45
		25	97.96
		50	110.96
5	Dhaka	10	79.41
		25	93.86
		50	104.58
6	Jessore	10	84.94
		25	106.14
		50	121.88
7	Sylhet	10	84.28
		25	100.03
		50	111.71

Source [3]

10.4   Storm Water–Disposal Guidelines

While draining and disposing of rainwater or stormwater from any building and its premises, some guidelines must be followed. The guidelines are as follows:

1. Rainwater from the roof or building premises should not be discharged into a septic tank. It should be drained into the storm sewer or combined sewer system, where available, or into surface drains leading to water course.
2. The water-seal trap must be installed on the building storm sewer, which should be connected to a combined sewer. The trap should be installed at the end of the building storm sewer inside the building site before it is connected to the combined sewer. No traps are required for building storm drains, which should be connected to the storm sewer for draining stormwater exclusively.
3. It is recommended to provide a secondary rainwater-drainage system at a suitable elevation from the roof, which should be considered in the calculation of rainwater load to design the building structure. The secondary drainage system should be a separate drainage piping leading up to the storm or a private

rainwater-disposal system. The size of the secondary rainwater-drainage piping should not be less than the size required for primary rainwater-drainage piping.

4. For the drainage of stormwater from unpaved surfaces, French drains may be employed as surface drains for this purpose. The construction of French drains should be in accordance with the established engineering code of practice. The location and alignment of these drains should be planned with respect to nearby all sources of waters, groundwater elevation, etc., and its design should be based on the area to be served and the maximum occupancy of the building.

10.5 Surface Drains

A rainwater or stormwater surface drain is a channel constructed on the ground surface to drain away stormwater quickly and efficiently toward a disposal destination. In areas of heavy rainfall or low soil percolation, it may be necessary to build surface-drainage systems for a building to eliminate surplus surface water from its various catchments and the surrounding land.

Surface drains can be developed in two ways: open-surface drain and covered-surface drain. Each system has advantages and disadvantages. The open-surface drain has the following disadvantages:

1. It receives various solid wastes, which creates a nuisance and reduces the flow capacity.
2. It is not hygienic and may emanate a foul odour.
3. Accidents may occur due to debris falling into the drain.
4. It may act as a breeding ground for mosquitoes and other organisms.

The advantages of an open-surface drain are as follows:

1. It is less prone to blockage.
2. It is easy to inspect and obtain access for removal of suspended and blocking wastes.
3. It is easy to identify the location and cause of the problem and rectify it accordingly.
4. Precautionary measures can be taken before problems occur.
5. It is easy to maintain and repair.

The development of effective surface drainage must be carefully planned and designed by taking into account the topography of the land and the peak runoff generated in the area to be served.

10.5.1 Computation of Peak Runoff

The rational formula used in computing the peak runoff from a catchment area is as follows.

$$Q_r = \frac{1}{360} C \times I \times A \qquad (10.2)$$

where

Q_r is the peak runoff at the point of design in m^3/s;
C is the runoff coefficient;
I is the average rainfall intensity in mm/h; and
A is the catchment area in hectares.

Runoff Coefficient: The runoff coefficient (C) depends on the degree and type of development within the catchments. Catchments are classified according to the expected general characteristics when fully developed. The C-values for various types of catchment areas developed are listed in Table 10.3.

Roughness Coefficient: The value of the roughness coefficient (n) depends on the type of material with which the surface of drain is finished. The roughness coefficient of various surface-finished materials is listed in Table 10.4.

10.5.2 Computation of Discharge Capacity

Drains are to be designed for steady uniform flow conditions wherein a one-dimensional method of analysis is considered. A drain should be designed in such a way that it must have adequate discharge capacities (Q_c) to accommodate the estimated peak runoffs (Q_r). The size, the cross-sectional geometry, and the gradient of a drain are the factors influencing its discharge capacity (Q_c). The calculated discharge capacity (Q_c) of a drain section must be at least equal to or preferably larger than the peak runoff (Q_r). The drain size is computed using Manning's formula given as below.

Table 10.3 C-values for various types of catchment areas developed

Characteristics of catchment when fully developed	C-value
Roads, highways, airport runways, paved areas	1.00
Urban areas fully and closely built up	0.90
Residential/industrial areas densely built up	0.80
Residential/industrial areas not densely built up	0.65
Rural areas with fish ponds and vegetable gardens	0.45

Source [2]

Note For catchments with composite land use or surface characteristics, a weighted C-value may be adopted

Table 10.4 The roughness coefficient of various finished materials of surface drains

Boundary condition	Roughness coefficient (n)
Unplasticised polyvinyl chloride (uPVC)	0.0125
Concrete	0.0150
Brick	0.0170
Earth	0.0270
Earth with stones and weeds	0.0350
Gravel	0.0300

Source [2]

Note Where there are different flow surfaces within a drain section, an equivalent roughness coefficient may be used

$$Q_c = \frac{1}{n} A \times R^{\frac{2}{3}} \times S^{\frac{1}{2}} \tag{10.3}$$

where

Q_c is the discharge capacity of drain (m^3/s);
n is the roughness coefficient of the drain-surface material;
A is the area in square metres from which the flow travels to the drain;
R $= A/P =$ hydraulic radius (m); where
P is the wetted perimeter (m); and
S is the gradient of the drain bed.

10.5.3   Surface-Drain Design Considerations

The surface drain should be designed adequately to provide passage for the conveyance of excess rainwater and stormwater based on climatic and soil conditions and the needs of the landscape. The design capacity of the drain should be based on the watershed area within the building premises and the grade line set at ground level. The design criteria are based on the following factors of flow in the drain:

1. Minimum velocity
2. Dry-weather flow
3. Maximum velocity and
4. Subcritical flow.

   Minimum Velocity: The velocity of flow in a drain is the most important factor to be satisfied for its effective functioning. The velocity of flow should must be maintained within a range of value suggested for different surface materials used in drain construction. The minimum flow velocity should be maintained so that the suspended matters present in the flowing rainwater or stormwater do not settle onto

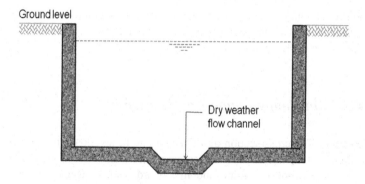

Fig. 10.1 Dry weather–flow channel in a surface drain

the drain bed. Therefore, this velocity is termed "self-cleansing velocity." The minimum velocity of flow in a concrete or masonry drain should not be <0.6 m/s for the self-cleansing action to take place. It should be noted that self-cleansing velocities increase with increasing drain size. Very large drains or sewers require high self-cleansing velocities. However, during dry seasons, the flow rate in a drain or sewer may fall to such a low level where this minimum velocity may not be achieved. To overcome this condition, a small channel is created in the drain or pipe bed to confine the dry-weather flow into that channel of smaller flow section, which is termed as the "dry-weather flow channel".

Dry-Weather Flow: To minimize siltation during low-flow conditions, a dry weather–flow channel is provided generally in the middle of the invert of the drain as shown in Fig. 10.1. It is a longitudinal shallow and narrow trough or channel of trapezoidal shape or a rectangular cross-section built as an integral part of a large flat-bottom surface of a main drain for concentrating the low flows to help develop self-cleansing velocities. A self-cleansing velocity of approximately 0.75–0.90 m/s [4] can be considered in sizing the dry weather–flow channel. If the dry weather flow fills the sewer or drain by >50 %, the slope that provides required self-cleansing at full flow under a velocity of 0.6 m/s can be used [5].

Maximum Velocity: The maximum velocity of flow in a drain should not be too high to cause scouring in channel bed and sides or a hydraulic jump in flow. Hence, the velocity of flow in a concrete and masonry–lined drain should be limited to a maximum of 3.0 m/s or less than the critical velocity, whichever is lower [2]. For non-vegetated open soil composed of loose rock or a heavy compacted clayey soil channel, the maximum velocity should be limited to 1.5 m/s; however, for extremely erodible soils it is only 0.3 m/s [6].

Subcritical Flow: Drains are generally designed considering that there will be subcritical flow in the channel. A critical-flow state exists when the Froude number is equal to 1.0. In an open channel, the flow of water at or near the critical state is generally avoided so that the water surface does not get unstable and wavy. To achieve greater efficiency in flow, the channel-flow conditions should be designed in such a way so that the Froude number falls as low as possible. It is preferred to have the Froude number within the range between 0.8 and the decreased minimum value, which allows to achieve practical flow depth and permissible flow velocity in the drainage flow [2].

10.5.4   Drain-Configuration Considerations

Cross-Section: The cross-section of a drain should be so chosen that it meets all of the requirements including flow capacity, velocity limitation, subcritical depth, stable side slopes, bottom width, and free board, and, if needed, allowances for sedimentation.

Side slopes should be designed based on soil conditions to make it stable against erosion. The drain cross-section should be set below the design hydraulic grade line taking care that the depth and width meet the maintenance requirements. A rectangular-shape surface drain, the width of which is greater compared with its depth, is more efficient. In some cases, a trapezoidal-shape drain, as shown in Fig. 10.2, or a V-shaped drain can also be designed and constructed.

The deposition of sediment, particularly sand and clay, in rainwater and stormwater drains is almost unavoidable. Therefore, the provision of sufficient allowance for the accumulation of sediments should be kept in the design criteria to determine the dimensions of the drain section. With a view toward accommodating the permissible degradation between de-silting cycles, thus decreasing the flow capacity due to materials deposited on the drain bed, the following guidelines are proposed [7]:

1. Five percent decrease in flow area if the gradient is >1 in 25.
2. Ten percent decrease in flow area in all other cases.

Depth: The depth of drains should be designed deep enough to accommodate normal siltation. Drains that will serve as outlets for subsurface main drains should be designed keeping its normal water surface at or lower than the invert of the outlet end of the main drain. The normal water surface is the elevation of the usual low-flow surface during the growing season. Where possible, the invert elevation of the main or lateral drain should be kept at least 300 mm lower than the invert elevation of subsurface drains that outlet into the main or lateral drains [8].

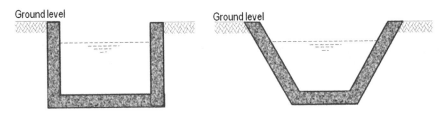

Fig. 10.2   Cross-sections of surface drains

Freeboard: Freeboard is the depth of the water surface in the drain at design-flow conditions from the top of the drain bank. Sufficient freeboard should be provided to avoid waves or fluctuation of the water surface from overflowing the drain bank. Generally, a depth of freeboard equivalent to 15 % of the depth of the drain is provided [2].

10.6   Subsurface Piped Drainage

The subsurface drainage system for storm or rainwater is for directing excess rainwater away from various catchments and surrounding land of buildings by natural or artificial means, generally by using a system of pipes placed below the ground-surface level. Underground storm-drainage pipes must satisfy two conditions: hydraulic and structural. Hydraulically these pipes must provide adequate passage for the estimated rainwater or stormwater for which they are designed. Equally important, structurally they must sustain the surrounding ground and support the over-burden of weight of the ground and the moving load on it if any.

10.6.1   Sizing of Stormwater–Drainage Piping

The size of building storm drain, storm sewer or any of their horizontal branches should be based on the maximum catchment area served including projected roof or paved area to be drained in accordance with Table 10.5.

Table 10.5 Size of horizontal-building storm drains and building storm sewer

Size of pipe (mm)	Maximum allowable horizontal projected roof areas (m²) for and flow (l/s) in horizontal rainwater drain piping at various slopes					
	10.4 mm/m		20.8 mm/m		41.7 mm/m	
	m^2	l/s	m^2	l/s	m^2	l/s
76	305	2.1	431	3.0	611	4.3
102	700	4.9	985	6.9	1400	9.8
127	1241	8.8	1754	12.4	2482	17.5
152	1988	14.0	2806	19.8	3976	28.1
203	4273	30.2	6057	42.7	8547	60.3
254	7692	54.3	10851	76.6	15390	108.6
305	12375	87.3	17465	123.2	24749	174.6
381	22110	156.0	31214	220.2	44220	312.0

Source [9]

Note 1 Table 10.5 is based on a maximum rainfall of 25.4 mm/h for a 1 h duration. The value for roof area of any locality should be determined by dividing the area given in the table by the desired rainfall in mm/h of that locality

2 The sizing data are based on considering the pipe to be running in full

10.6.2   Structural Safety of Stormwater-Drainage Piping

The forces acting on a cross-section of a subsurface rigid pipeline arise from three main sources as follows:

1. The weight of overlying fill including any dead or live surcharge load
2. Traffic and other transient loads on the surface transmitted as soil pressure on the pipe
3. Supporting reaction of soil below the pipe.

In general, pipelines are laid in trenches and then covered by backfilling. The backfilling structures may be built subject to possessing the structural capacity of the pipe and the soil capacity below. Therefore, the buried pipes are to be designed to make them capable to withstand the backfill load, traffic loads, and—when the diameter is ≥600 mm—part of the water load within the pipe. In determining the structural safety of the buried drainage pipe, the following factors must be considered

Bedding class and factors: The load-carrying capacity of a concrete pipeline is dependent on both the flexural strength of the manufactured pipe and the support provided by the bedding developed. "Bedding factor" (F_m) is the term used to denote the ratio of the strength of the laid pipe to its laboratory-based crushing strength. The higher the bedding factor, the greater the load-bearing capacity of a given concrete pipeline. The bedding-factor values for different classes of bedding are listed in Table 10.6.

Laboratory-crushing strength: The crushing load (W_t) of a concrete pipe is the minimum load that the pipe will sustain without collapse. Crushing test loads in kiloNewtons per metre (kN/m) of effective length for concrete pipes of different diameter are listed in Table 10.7

Safe-supporting strength: For an underground buried pipe, the safe-supporting strength must be checked. The safe-supporting strength of a pipe must be greater than all of the external pressures working on the pipe. The supporting strength of a pipe is the product of the pipe-crushing strength and the bedding factor. The total external load acting on pipe is the summation of the backfill load, the surcharge or traffic load, and—for pipes greater than DN600—the equivalent water load.

Table 10.6 Bedding-factor values for different classes of bedding

Sl. No.	Bedding class	Bedding description	Bedding factor
1	D	Flat subgrade	1.1 [10]
2	F	45° granular bed	1.5 [11]
3	B	180° granular bed	1.9 [10]
4	S	360° granular bed	2.2 [11]
5	A-nonreinforced	120° nonreinforced concrete cradle	2.6 [11]
6	A-reinforced	120° reinforced concrete cradle	3.4 [11]

Source [10, 11]

Table 10.7 Minimum crush-test loads for concrete pipe as stated in I.S. 6: 2004

Type of concrete pipe	Nominal pipe φ (DN) in mm	Minimum crushing load Fn kN/m (Fn = weight)
Unreinforced or plain concrete pipe	225	27
	300	36
	375	45
	450	54
	525	63
	600	72
	675	81
Reinforced concrete pipe	750	90
	900	108
	1050	126
	1200	144
	1350	162
	1500	180
	1650	198
	1800	216
	2100	252

Source [12]

The crushing strength of a concrete pipe can be determined using the following formula [12]:

$$\text{Minimum crushing strength} \quad W_t > W_e \times F_s / F_m \tag{10.4}$$

where

W_e is the total external load (kN/m);
F_s is the factor or safety taken as a minimum of 1.25 for concrete pipe and 1.5 for reinforced concrete pipe [11]; and
F_m is the bedding factor.

10.7 Grading of Land Surface

The purpose of grading the land surface is to provide natural and gentle surface drainage for the stormwater. For having complete and adequate drainage on the ground, land grading is required where the surface is undulated and has reverse surface grades forming depressions. The recommended surface grades range from 0.1 to 0.5 % [13]. Surface grading may be uniform or variable. The cross-slopes should not exceed 0.5 % [13].

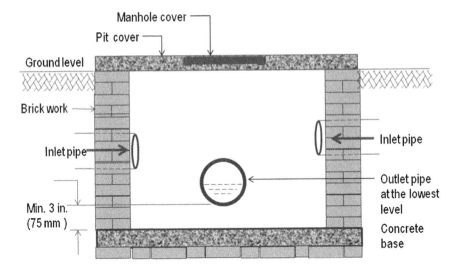

Fig. 10.3 Section of an inspection pit

Land grading is also termed "precision land forming". It is generally created by reshaping the undulated land surface with the help of tractors and scrapers and finally finished manually to a planned grade. To plan and make a good land grading, a detailed engineering survey to obtain the contour of surface of land is required.

Land grading is hampered by trash and vegetation the growth of which should be controlled or removed before construction. The land surface should be firm and dry when it is surveyed.

10.8   Manholes or Inspection Pits

A manhole or inspection chamber is required in stormwater drainage at specific locations to obtain access to the drainage system for unblocking, cleaning, rodding, and inspecting the flow condition.

The term inspection "chamber" or "pit" is used when the depth to the drain is ≤1 m. For drains lying deeper, more robust and deeper chambers are required depending on the size and number of drain pipes to be accommodated; these are termed "manholes." Figure 10.3 illustrates the detail of an inspection pit used in stormwater drainage.

10.8.1 Location

Inspection pits or manholes on underground piped drains should be provided at the following locations.

1. Intersections of stormwater drains
2. Junctions between different sizes of stormwater drains
3. Where a stormwater drain changes size, direction, and gradient
4. On long straight lengths at certain intervals as listed in Table 10.8

In some special cases, as mentioned below, for pipe of size ≤675 mm, the interval of manholes or inspection pits can be decreased to 60 m, as follows:

1. There is possibility of frequent blockage in the flow.
2. The opening of adjacent manhole covers at the same time may cause difficulties in traffic or pedestrian movement.
3. Pits are located in a narrow roads that are inaccessible to standard water-jetting units.

10.8.2 Size of Inspection Pits or Manholes

An inspection pit or manhole chamber can be designed to be round, square, or rectangular. Round manholes, as shown in Fig. 10.4, are the most widely used for structural reasons when the depth is greater. The dimensions of a chamber or manhole should be determined by considering the number, size, and alignment of the branch pipes or spurs feeding into it. Inspection chambers are made to be rectangular or circular having a minimum internal dimension of 450 mm and a maximum depth considered to be 1000 mm. The minimum internal dimensions of a manhole should be 750 by 1200 mm for a maximum depth of maximum 2.7 m [15]. Circular manholes are commonly used for a main drain; for depths ≤1.5 m, they must have a minimum diameter of 1050 mm; for anything deeper than 1.5 m, the diameter must be 1200 mm [15].

At particular junctions, the size of the manhole or inspection pit depends mainly on the number and size of the pipes to be accommodated. While connecting larger-sized and a greater number of pipes in a manhole, there is chance of

Table 10.8 Intervals of inspection pits on long straight sewers

Diameter of pipe (mm)	Maximum intervals (m)
≤675	80
>675–1050	100
>1050	120

Source [14]

Fig. 10.4 Plan of a circular
manhole

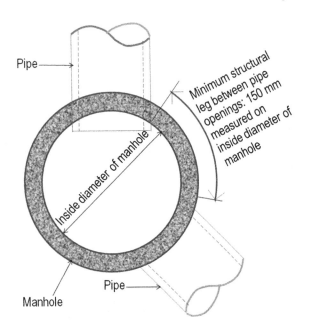

impairing the structural strength of the manhole. Therefore, manhole sizing includes
the number and the outer diameter or dimensions of the pipes at approximately the
same invert level; the thickness of its walls; and the distance, or leg, between pipe
openings. The criteria to be followed in sizing manholes are as follows:

1. Manholes must be large enough to accommodate the maximum intersecting pipe
   size or the maximum number of pipes at approximately the same invert level.
2. Between two adjacent pipe openings in a manhole, the minimum leg, or
   structural length, must be maintained.
3. As a general rule based on the outside diameter of the intersecting pipe, the
   minimum concrete leg between openings for adjacent pipes in manholes is
   150 mm [16].
4. Block-out or cut-out diameters are based on the outside diameter of the pipe plus
   a distance not to exceed 100 mm [16].

For manholes, the minimum pipe opening is assumed to be the inside diameter
of the pipe plus its wall thickness. The minimum structural leg, as a practical
industry guideline, is taken as 150 mm [16].

The depth of a manhole is primarily governed by the invert of the deepest outlet
pipe, which again is placed approximately 6–12 mm lower than the deepest inlet
pipe in the manhole. Below the invert of an outlet pipe, a depth of approximately
75–150 mm is provided to create a chamber for the accumulation of silt carried by
stormwater or rainwater.

References

1. Sarala C, SreeLakshmi G (2014) Improvement of storm water drainage system in greater Hyderabad municipal corporation. Int J Adv Eng Technol, 7(2): 605–613, 608
2. PUB Singapore's National Water Agency (2011) Code of practice on surface water drainage, 6th edn., Retrieved on 07 Dec 2015 http://www.pub.gov.sg/general/code/Documents/HeadCOPFINALDec2011-1.pdf
3. Un-named (Un-dated) Dissertation of post graduate diploma in civil engineering under joint program of Asian Institute of Technology (AIT) Bangkok, Thailand and Bangladesh University of Engineering and Technology (BUET), Dhaka, p 26
4. Fair GM, Geyer JC, Okun DA (1971) Elements of water supply and wastewater disposal, 2nd edn., John Wiley & Sons Inc. and Toppan Company Ltd. Cited in 'Wastewater and storm water collection and removal' by Bertuğ Akıntuğ, Middle East Technical University. Retrieved on 10 Dec 2015 http://users.metu.edu.tr/bertug/CVE471/CVE%20471%20-%206%20Wastewater%20and%20Stormwater%20Collections%20and%20Removal.pdf
5. Akıntuğ B (2015) Wastewater and storm water collection and removal, water resources engineering. Middle East Technical University, Northern Cyprus. Retrieved on 10 Dec 2015 http://users.metu.edu.tr/bertug/CVE471/CVE%20471%20-%206%20Wastewater%20and%20Stormwater%20Collections%20and%20Removal.pdf
6. Brisbane City Council, Natural Channel Design Guidelines. Retrieved on 7 Dec 2015 http://www.brisbane.qld.gov.au/sites/default/files/ncd_sect3_design_part7.pdf
7. Government of the Hong Kong (2013) Storm water drainage manual, planning, design and management, 4th edn., p 51, Drainage services department. Retrieved on 10 Dec 2015 http://www.dsd.gov.hk/EN/Files/Technical_Manual/technical_manuals/Stormwater_Drainage_Manual_Eurocodes.pdf
8. United State Department of Agriculture (USDA) (2015) Surface drain, main or lateral conservation practice standard, code 608. Natural Resources Conservation Service
9. Plumbing Code IAPMO, Standards India (2007) Uniform plumbing code- India 2008. IAPMO Plumbing Code and Standards Pvt Ltd., India, p 170
10. Ontario Concrete Pipe Association (OCPA) Concrete pipe design manual. Retrieved on 27 Oct 2015 http://www.ocpa.com/_resources/OCPA_DesignManual.pdf
11. CPM Group, CPM concrete drainage systems Pipeline design—Hydraulic, UK, www.cpm-group.com
12. Condron Concrete Works, Concrete pipes, Brocheur. Retrieved on 27 Oct 2015 http://www.condronconcrete.com/pipenew_brochure.pdf
13. University of Illinois, Surface drainage, illinois drainage guide (Online), Department of Agricultural and Biological Engineering. Retrieved on 27 Oct 2015 http://www.wq.illinois.edu/dg/surface.htm
14. Government of Hong Kong (2013) Sewerage manual, key planning issues and gravity collection system, 3rd edn., p 39. Retrieved on 10 Dec 2015 http://www.dsd.gov.hk/EN/Files/Technical_Manual/technical_manuals/Sewerage_Manual_1.pdf
15. www.pavingexpert.com, Manholes and inspection chambers. Retrieved on 27 Oct 2015 http://www.pavingexpert.com/drain05.htm
16. National Precast Concrete Association (NPCA) (2010) Tech topic: manhole sizing recommendations contributed by Claude Goguen. Retrieved on 10 Dec 2015 http://precast.org/2010/11/tech-topic-manhole-sizing-recommendations/

Chapter 11
Rainwater-Harvesting Elements

Abstract Rainwater harvesting in buildings involves various items and works of multidisciplinary technology and materials. It may not be possible to acquire multidisciplinary technological knowledge by every professional or decision maker involved in rainwater harvesting. Therefore, for professionals and other practitioners engaged in rainwater harvesting in buildings, knowledge about various items of different technologies is essential for their effective and efficient utilization. It may not be possible to acquire in-depth knowledge of all of the types of elements used; however, some basic concepts are essential.

This chapter is designed to introduce all of the elements and the characteristics related to rainwater-harvesting systems in buildings and their premises. The basic concepts of these elements are furnished herein for helping the sustainable development of a rainwater-harvesting system in buildings.

11.1 Introduction

Various elements used in rainwater harvesting in buildings are for performing various jobs. All these elements used are found in various types, natures, and qualities. Therefore, to select the appropriate element, various aspects of those elements should be known. In this chapter, the very important aspects of those various elements related to rainwater harvesting in buildings are discussed.

11.2 Elements in Rainwater Harvesting

It has already been mentioned that there are two broad perspectives of rainwater harvesting. Both systems have common elements. The elements found in both of these systems are as follows:

© Springer International Publishing Switzerland 2017

S.A. Haq, PEng, *Harvesting Rainwater from Buildings*,

DOI 10.1007/978-3-319-46362-9_11

A. Elements involved in general-purpose uses of rainwater are as follows:

1. Pipe
2. Storage tank
3. Catchments
4. Pump
5. Valve and
6. Faucet.

B. Elements involve in groundwater recharging are as follows:

1. Pipe
2. Filter media and
3. Recharge structures

   (a) Brick
   (b) Sand
   (c) Cement
   (d) Water
   (e) Gravel
   (f) Boulder
   (g) Reinforcing bars and
   (h) Manhole cover.

11.3 Pipe

Pipe is the major component of rainwater harvesting, which is a cylindrical-shaped element used for conveying fluids; here rainwater and stormwater are the only concerns. Various types and different classes of pipes are manufactured to meet the varied condition of service. Generally three classes of pipes based on materials are used in rainwater harvesting. These are as follows:

1. Galvanized iron (GI) pipe
2. Plastic pipe
3. Cast iron pipe and
4. Concrete pipe.

11.3.1 Galvanized Iron Pipe

Galvanized iron (GI) pipes are iron pipes with a galvanized coating. Galvanizing is done by dipping clean-surfaced iron pipes into molten pure zinc. Galvanizing on and inside of iron pipes is performed mainly to prevent pipes from corroding. GI pipes are manufactured in three classes: class A (light), class B (medium), and class C (heavy). The allowable working pressure for a class-B GI pipe is 20 kg/cm^2 and for a class-C pipe is 30 kg/cm^2 [1]. GI pipes are connected mostly by threaded joints. Welding of galvanized pipe should be avoided because it results in the emission of a toxic gas from the zinc. If the rainwater is corrosive due to its acidic condition (low pH), the leaching of lead and cadmium, which may be present as impurities in the zinc, may occur.

11.3.2 Plastic Pipe

Plastics are a family of man-made materials developed from synthetic organic chemicals such as oil, natural gas, coal, and cellulose (from wood fibers). These raw materials are made into resins. Tests have indicated that plastic pipe may last for <50 years [1] under favorable conditions, which is darkness. The advantages and disadvantages of using plastic pipe are as follows:

1. Tuberculation cannot occur due to there being no deposition of soluble encrustants such as $CaCO_3$.
2. Plastic has good resistance to biological attack.
3. Plastic is resistant to almost all types of corrosion.
4. Plastic is comparatively light in weight.

The disadvantages of using plastic pipe are as follows:

1. Plastic pipe is damaged by ultraviolet radiation from sunlight.
2. The performance of plastic pipe degrades with the increase of both internal and external temperature.
3. Plastic pipe is subject to permeation by pollutants of the lower molecular weight of organic solvents or petroleum products.
4. Plastics pipe is comparatively not very long-lasting.
5. Plastic pipe undergoes an aging effect, and thus the quality degrades with time.

Types of plastic pipe: Various types of plastic pipe can be used in rainwater harvesting in buildings. The common plastic pipes used in rainwater supply or distribution are unplastisized polyvinyl chloride (uPVC) pipe and chlorinated polyvinyl chloride (CPVC) pipe. In rainwater drainage, another type of plastic pipe should be used, which is called "acrylonitrile butadiene styrene" (ABS) pipe.

uPVC pipe: uPVC pipe is used for normal rainwater supply; therefore, uPVC pipe can be used for all purposes of rainwater use and groundwater recharging. These pipes are manufactured of plastic resins extruded in special machines under a closed-temperature control-and-manufacturing process. uPVC pipes are suitable for operating under temperatures from 0 to 60 °C [2]. These pipes should not be installed exposed to direct sunlight due to its high coefficient of thermal expansion, which is approximately 0.08 mm/m/°C [2], as well as loss of strength. Rigidly fixed pipes may deform due to high thermal expansion. As a result, a joint may start leaking. Therefore, provision for pipe expansion should be maintained in the piping system.

CPVC pipe: CPVC pipes are useful for handling high temperature, corrosive fluids, having a maximum service temperature up to 99 °C [3]. It has good chemical resistance like uPVC pipes. If hot water is required in any building, hot rainwater should be supplied there using these pipes when plastic pipe is chosen for this purpose. These pipes may also be used in cold water lines.

ABS pipe: ABS pipe has a high impact strength, is very tough, and can be used at temperatures ≤70 °C [4]. It has lower chemical resistance and lower design strength than uPVC pipe. This type of pipe is generally used in conveying water for irrigation, gas transmission, draining wastes, and venting. ABS pipe can be used in draining stormwater. Therefore, solvent welding and threading are the recommended process for jointing these pipes.

Plastic pipe remaining exposed to direct sunlight will degrade over time, thus becoming brittle and weak; therefore, they can suddenly break or cause damage without any previous notice. After a rainwater-harvesting system is installed, plastic pipes should be painted. The first step in painting should clean the pipe with a degreaser such as trisodium phosphate (TSP). Next, a primer that works on plastic should be applied. After application of primer, the pipe should be painted with a water-based paint. Painted plastic pipes will prevent UV light from reaching the bare pipe, thus retarding degradation over time. Therefore, some paints are specifically made for use on plastic by spraying it directly on the pipe surface. These types of paints dry quickly.

11.3.3 Cast Iron Pipe

Cast iron (CI) pipes are manufactured from gray pig iron. Cast iron contains a significantly greater percentage of carbon and silicon. These pipes are resistant to corrosion and have a long life of approximately 100 years [5]. Cast iron pipes of socket and spigot type are available in various diameters ranging from 80 to 1050 mm for an effective length of 3–6 m [6]. The major disadvantages of these pipes are their very poor or nonelastic behavior, lower mechanical strength, and propensity toward external and internal corrosion under aggressive conditions. To protect from surface corrosion, paint should be applied to the external surface and cement mortar lining to the internal surface. Cast iron pipes are manufactured in four different forms as shown in Fig. 11.1. Cast iron fittings are also manufactured for use in this type of piping system.

Fig. 11.1 Various forms of cast iron (CI) pipe

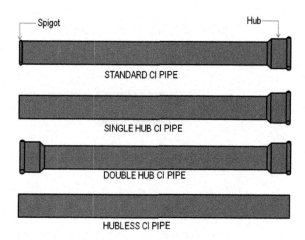

11.3.4 Concrete Pipe

Concrete pipe is supposed to be the most durable, long-lasting, and cost-effective pipe amongst all other pipes for underground use. Considering strength, concrete pipes can be classified as (1) nonpressure pipe, (2) pressure pipe, and (3) special-purpose pipe. Reinforced concrete pipe is a composite structure and is specially designed using reinforcement in cement concrete. Concrete pipe may be nonreinforced or reinforced. To make reinforced cement concrete (RCC) pipe, iron wire or steel bars are used as reinforcement. With these reinforcing wires or bars, cylindrical-shaped cages are made, which are finally embedded in cement concrete to shape up as pipe. Although concrete pipes can be manufactured in various shapes, mostly circular pipes are used for their good self-centering and efficient hydraulic characteristics [7]. The diameter of nonreinforced concrete pipe generally manufactured ranges from 100 to 900 mm; reinforced concrete pipes are manufactured ranging in diameter from 150 to 3600 mm [7]. The crushing strength for concrete pipes of different diameter is given in Chap. 10.

Concrete pipe ends are generally constructed in three forms for different type of jointing as shown in Fig. 11.2.

1. Bell-spigot type
2. Rebated joint (tongue-and-groove) type and
3. Butt joint (plain-end) type.

Concrete pipes are manufactured in effective lengths of 1.2, 2.0 and 2.4 m. For special cases, 0.75 m-long pipes are also manufactured.

The main advantage of concrete pipes is their corrosion resistance. The disadvantages are that these pipes are bulky and heavy and thus require careful transportation and handling. The layout process of these pipes is costlier than that of plastic or iron pipes. The life of concrete pipe is generally considered as long as ≥100 years [8].

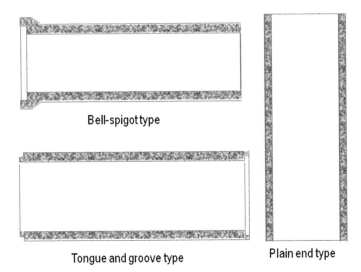

Fig. 11.2 Concrete pipes of various jointing systems

11.4   Pipe Fittings

Pipe fittings are devices attached to the pipe for following purposes.

1. To alter the direction and diameter of pipe
2. To make branching in pipe
3. To connect two pipes, pipe with fixtures, accessories, or equipment and
4. To close an end of pipe.

Pipe fittings are also grouped and classified with respect to their various uses as follows.

11.4.1   Fittings for Jointing

For jointing of pipes of various materials, the following fittings, made of respective material, are used.

Nipple: Nipples are used for jointing two pipes or fittings. These are generally provided with a male pipe thread at each end. Various sizes of nipples are available as close, short, and long as shown in Fig. 11.3.

Coupling: Coupling or extension pieces are used for connecting two pipes to each other. These are generally provided with a female pipe thread at each end as shown in Fig. 11.4.

Unions: A union is a piece of fitting equipment, as shown in Fig. 11.5, that joins two pipes of the same size. Union is designed in such a way that allows quick and

Fig. 11.3 Various nipples

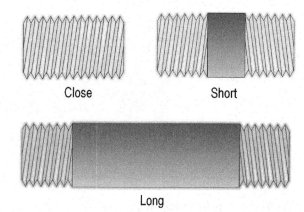

Fig. 11.4 Coupling

Fig. 11.5 Union

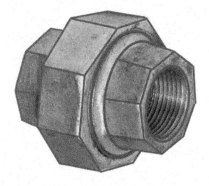

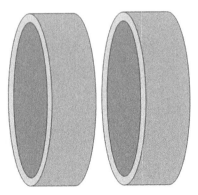

Fig. 11.6 Collars for jointing RCC pipes

convenient disconnection of pipes for maintenance or fixture replacement without causing damage or deformation in pipes.

Collar: A collar, as shown in Fig. 11.6, is a ring-type element made of reinforced cement concrete (RCC) that is used for jointing RCC pipes used mostly in drainage.

11.4.2   Fittings for Changing Pipe Diameter

For decreasing or enlarging the diameter of pipes, the following fittings are used.

Reducers: Reducers are used to join pipe or any accessories of decreased size. Reducers are again of two types, such as concentric reducer and eccentric reducer, as shown in Fig. 11.7.

Bush: A bush is used for joining smaller-diameter accessories on a larger-diameter pipe. A bush should not be used for decreasing the pipe diameter in any fluid supply. It posses both male and female threads as shown in Fig. 11.8.

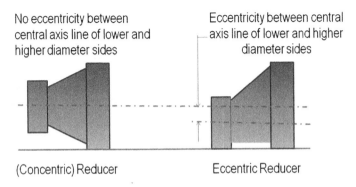

Fig. 11.7 Various reducers

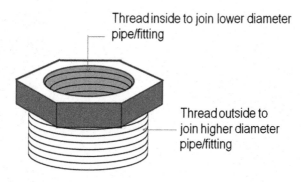

Thread inside to join lower diameter pipe/fitting

Thread outside to join higher diameter pipe/fitting

Fig. 11.8 Bush

11.4.3   Fittings for Changing Direction

For changing the direction a pipe runs, the following fittings are used.

Elbow: Elbows are used for the changing direction in which a pipe runs. For changing the direction of a pipe run at various angles, elbows of different angles (as shown in Fig. 11.9) are designed. Mostly 90° elbows are used. Long-sweep elbows cause comparatively less frictional loss. In drainage, the use of 45° elbows is preferred.

Offset: An offset, as shown in Fig. 11.10, is used to shift the axis of a pipe run.

11.4.4   Fittings for Pipe Branching

To achieve branching on pipes, the following fittings are commonly used.

Tee: A tee is used to make a branching on pipe at a 90° angle. Figure 11.11 represents a tee fitting.

Cross: A cross is used to make two pipes branch from the same location in opposite directions. Figure 11.12 shows a cross fitting.

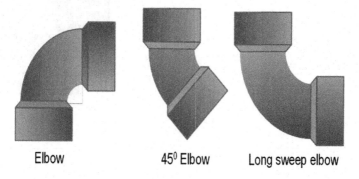

Elbow   45⁰ Elbow   Long sweep elbow

Fig. 11.9 Various elbows

Fig. 11.10 An offset

Fig. 11.11 Tee

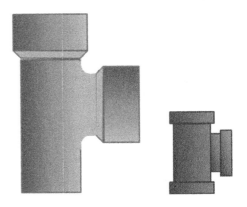

Fig. 11.12 Cross

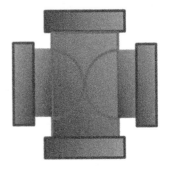

11.4.5   Fittings for Pipe Closing

For shutting off or closing a pipe end, the following fittings are commonly used.

Plug: A plug is used inside a pipe end to close it. Plugs have male thread as shown in Fig. 11.13.

Cap: A cap is used on a pipe end to close it. A cap has an internal thread as shown in Fig. 11.14.

Flange: A flange is used to close a pipe end using bolts as shown in Fig. 11.15.

Fig. 11.13 Plug

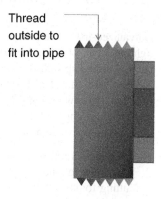

Thread outside to fit into pipe

Fig. 11.14 Cap

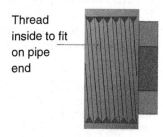

Thread inside to fit on pipe end

Fig. 11.15 Flange

11.5   Pipe Jointing

Pipe jointing is the most vital job in achieving a good plumbing system for rain-water harvesting in buildings. Methodologies for pipe jointing are different for pipes of different materials. The jointing process for pipes of different materials is discussed below.

11.5.1   GI Pipe Jointing

Galvanized-iron pipe joints are generally made a taper-thread joint for making it tight and leak-proof. Of various type of taper threading, there are two widely used tapered thread specifications in practice: the American Standard Taper Pipe Thread (NPTF) and the British Standard Pipe Tapered (BSPT). The basic difference between these two standards is shown in Fig. 11.16. The tapering of thread is made usually at an angle of 1.47°.

The process of jointing GI pipe should be followed as mentioned below.

1. The pipe should be cut square by properly securing the pipe in the vise jaw.
2. The inside of the cut should be reamed to remove the burr often left by the cutting tool.
3. The pipe is threaded with a die to the proper length while lubricating the die.
4. After threading, pipe is removed from the die and the thread is wiped clean.
5. The pipe joint compound is applied to the male threads only.
6. The fitting is twisted onto the pipe thread first by hand, and then the joint is tightened with the help of a pipe wrench.

Fig. 11.16 Pipe-thread tapering according to both American and British standards

(**a**)   American Standard Taper Pipe Thread (NPTF)

(**b**)   British Standard Pipe Tapered (BSPT).

11.5.2 Plastic-Pipe Jointing

Plastic pipe and fittings are bonded together by the process of chemical fusion. Various solvents are used as a primer or for cementing purposes in jointing plastic pipes and fittings. Solvent contained in primer and solvent cement virtually softens and dissolves the surfaces to create a chemical reaction between the surfaces to be jointed. Once the solvent is applied, the surfaces of the pipes or fittings are assembled together, and a chemical weld occurs between the surfaces. This chemical welding gains strength over time as the solvents evaporate. The process to be followed in making a perfect plastic jointing is mentioned below and shown in Fig. 11.17.

1. Plastic pipe should be cut squarely using a meter box or a sharp tube cutter with special blade for plastics.
2. By using a debarring tool or knife, all burrs of a cut surface should be removed to make it smooth.
3. The pipe end should be prepared by rubbing it with sand paper for a good interface contact.
4. The jointing surfaces of the pipe and fittings should be cleaned and dried.
5. An appropriate primer should be applied first to the inside of the fittings and then to the outside of the pipe end portion, which will remain inserted into the fitting.
6. The pipe should be inserted into the socket immediately before the solvent cement evaporates. While pushing, a quarter turn of the pipe should be made. Next the joint should be held together firmly for approximately 10 s.
7. Finally the excess solvent cement should be wiped off with a clean rag.

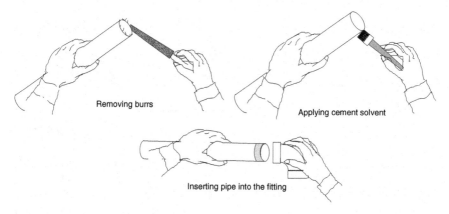

Removing burrs

Applying cement solvent

Inserting pipe into the fitting

Fig. 11.17 Plastic pipe-jointing process

11.5.3 Cast-Iron Pipe Jointing

In cast-iron pipe jointing, generally lead is used as a jointing material. To make a strong and leak-proof joint when jointing a cast-iron pipe, the following procedures are carried out.

1. The hub (bell) and spigot ends of the pipes are cleaned and dried.
2. The pipes are assembled by inserting the spigot end centering into the bell to its full depth as shown in Fig. 11.18.
3. Yarn or oakum is then packed into the annular space around the pipe inserted into the bell to a depth of 25 mm from the top of bell.
4. In vertical pipe jointing, using a ladle, molten lead is poured into the remaining portion of the annular space. The space should be filled in one pour. In horizontal pipe jointing, in order to facilitate the pouring of lead, an asbestos running rope is placed around the pipe and clamped tightly at the top, thus forming a passage for the lead to fill the annular space. A wad of oakum is placed under the clamp to retain the lead up to the top of the bell. The lead is then poured into the passage, thus filling the joint to the top. The running rope is removed when the lead solidifies.
5. As the molten lead hardens, it is caulked with light taps to compensate for shrinkage of the lead as it cools. The neck of excess lead left in the pouring access is cut off and finished beveled with a chisel and hammer.

Fig. 11.18 Cast-iron pipe joint

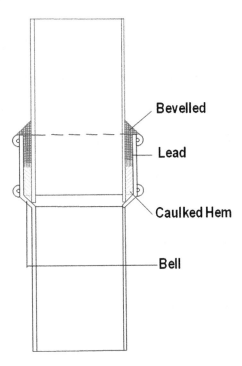

11.5.4 Concrete-Pipe Jointing

Concrete pipes are generally jointed in two methods: a spigot-and-socket joint and a butt joint with collar.

Spigot-and-socket joint: Concrete pipes joined in this method have external shoulders formed to create a bell at one end, whereas the other end has the spigot formed with a notch to house an o-ring or other rubber gasket as shown in Fig. 11.19. In the absence of a rubber gasket, a jute oakum approximately 19-mm thick is caulked into the bell. The remaining annular space is filled by cement mortar prepared by using one part cement, two parts sand, and sufficient water to dampen thoroughly to such a consistency that it prevents grout from running into the pipe. This type of joint is recommended in soft soil and a water-tight joint under low-head applications.

Butt joint with collar: This type of joint uses a precast or cast in situ concrete collar to connect the pipes. When jointing, the pipe end are aligned face-to-face and put just in the middle of the collar. The annular space between the collar and the pipe is filled with cement mortar. Small-diameter concrete pipes are relatively easy to line up; this type of joint can be created easily. This is a rigid joint, which does not allow any deflection. For this reason, such a joint is not recommended for soft soil where there might be a deflection in pipe resulting in cracks in the joint.

11.5.5 Pipe-Joint Testing

To make a leak-proof joint and to achieve durability, pipe joints should be tested for ensuring its capability of withstanding the anticipated hydraulic pressure to be exerted at the joints. Various methodologies can be adopted for this testing. The most practiced methodology of testing pipe joints is discussed below.

Water Test: This methodology of testing joints is generally applied to pipes conveying water. The testing procedure is as follows.

Fig. 11.19 Concrete-pipe joints

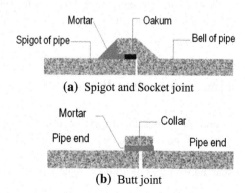

(a) Spigot and Socket joint

(b) Butt joint

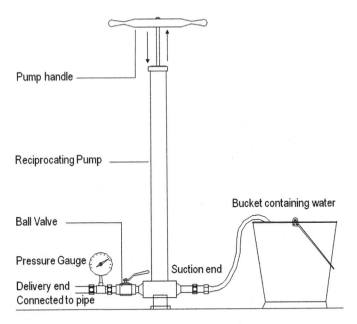

Fig. 11.20 Pressure-developing appliances for pipe-joint testing

1. Install bib cocks at the two ends of the piping system to be tested: one for filling water into the piping and the other for evacuating air from the piping.
2. Put on the cap and make watertight all openings kept for connecting faucets or fixtures with the piping.
3. Fill the piping with potable water through one bib cock, keeping the other bib cock open, until the system is full of water and inside air is driven out fully. The air-venting pipe end is then capped tightly.
4. Increase the pressure in the system by a hand pump, as shown in Fig. 11.20, so that it is not less than the following: (1) 150 % of the working pressure at the point of test and (2) however, not less than 125 % of normal working pressure at the highest elevation under which the piping system will be operated [9].
5. Sustain the pressure with the system for approximately two hours [9]. If the pressure remains the same, then the system is proved to be watertight.

11.6   Supporting Pipes

Supporting pipes at proper locations is another important job to be performed. Various pipe supports are available to support pipes of different sizes and classes for installing at different locations as shown in Fig. 11.21. Recommended pipe support spacing for pipes of different materials and various diameters is listed in Table 11.1.

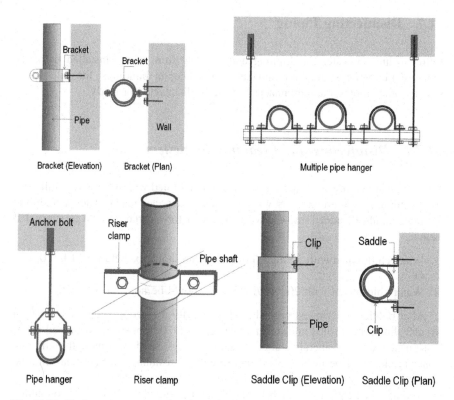

Fig. 11.21 Various pipe supports for pipes of different sizes and classes. Modified from source [11]

Table 11.1 Pipe-support spacing for pipes of different materials and various diameters

Sl. no.	Piping material	Maximum horizontal spacing (m)	Maximum vertical spacing (m)
1	Galvanized steel pipe	3.5	4.5
2	Copper pipe or tube a. Smaller than 38 mm diameter b. larger than 38 mm diameter	3.5 2.0	3.0 3.0
3	PVC pipe and tube	1.0	1.0
4	Aluminium tube	3.0	4.5
5	Brass pipe	3.0	3.0

Source [10]

11.7   Disinfecting Piping

Before using rainwater for drinking and cooking from newly installed or partly replaced plumbing system for rainwater harvesting in buildings, the full piping system—including all appurtenances—must be disinfected properly.

11.7.1   Disinfecting and Cleaning the System

First, the whole rainwater-supply system should be filled with water to the full brim of the topmost appurtenant or wherever the water can reach. Then a chlorine solution should be added into the water and kept for a certain period as suggested below.

1. Water chlorinated with 50 mg/l (ppm) chlorine should be held for 24 h [10, 12]. Or
2. Water chlorinated with 200 mg/l chlorine should be hold for 3 h [10].

After holding chlorinated water in the system for the period suggested, the total system should be flushed until residual chlorine of 0.02 ppm [12] is attained in the last-flushed water. Generally, after flushing the system three times, the chlorine concentration will be decreased to the acceptable residual chlorine concentration limit.

Adding 24.12 g (by weight) of bleaching powder ($CaOCl_2$) containing approximately 75 % chlorine to 1000 gallons of water produces a chlorine concentration of 1 ppm. Therefore, for a 50-mg/l chlorine concentration, approximately 1.20 kg of bleaching powder should be added to 1000 gallons of water, and for 200-mg/l concentration, approximately 4.80 kg of bleaching powder should be added to 1000 gallons of water at pH 7.5 and temperature 25 °C.

11.8   Storage Tanks

Storage tanks are other important elements of rainwater harvesting in buildings. Storage tanks are either constructed or assembled on-site or directly installed. Generally, commercially available smaller-sized tanks (particularly in volume) made of various materials are directly installed. Masonry or reinforced cement concrete tanks of comparatively larger in size are constructed on the site where they are to be located. Larger tanks made of iron or steel sheets are generally assembled at the location site. The selection criteria for storage-tank materials should be based on economy, durability, and possibility of water contamination. Tanks of various materials are discussed below.

11.8.1 RCC or Masonry Tanks

Constructed tanks on the ground or roof, generally with capacities exceeding 50,000 l, are mostly made of reinforced cement concrete (RCC). The RCC tank is less expensive for tanks sized ≤1,000,000-l capacity. For tanks >1,000,000-l capacity, pre-stressed concrete tank is supposed to be less expensive by approximately 20 % [13] compared with the cost of an RCC tank of that size. Walls of underground of those with smaller depth can be made of bricks. Generally tanks of capacities ranging from 15,000 to 50,000 l can be made of bricks [14]. Tanks can be built in any shape and size with these materials. Masonry tanks are economic, however, making them water-tight is difficult.

11.8.2 Ferro-cement Tank

Ferro-cement is a cement-based composite construction material; it is modified form normal reinforced cement concrete by using mesh reinforcement instead of reinforcing bars. This type of construction is an effective and durable construction material for water tanks. Ferro-cement water tanks can be used for rainwater storage in buildings. A commercially available ferro-cement tank of 1.2 m^3 area, as shown in Fig. 11.22, is built by assembling 20 numbers of 525 mm^2 ferro-cement plates of 12-mm thickness made of cement and sand reinforced with two layers of 18 BWG wire mesh [15]. Ferro-cement tanks can be found pre-assembled and also in parts for erection at the site. Techniques for constructing ferro-cement tanks of ≤25,000-l capacity have been developed [16].

Fig. 11.22 Ferro-cement tank

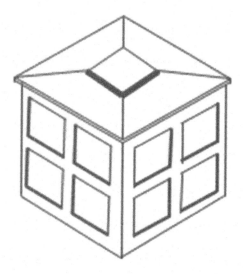

If rainfall is acidic and contains very few mineral salts, the harvested rainwater stored in an RCC, masonry, or ferro-cement tank becomes neutral or lightly alkaline (pH between 7.5 and 8.5) and is also weakly mineralized.

11.8.3   G.I. Tank

These tanks are fabricated of galvanized-iron sheets. Smaller tanks can be made of 16–18 BWG-thick iron sheets by riveting with GI rivets at corner edges. The maximum capacity of these tanks is ≤1800 l. Corrugated galvanized iron (CGI) tanks are also available, which are manufactured of corrugated iron or steel sheets of 20-gauge thickness [17]. With this type of sheet, tanks of ≤300,000-l capacity can be built [17]. This type of metal tank is popular for aesthetic reasons, fire resistance, unchanged quality of stored water, and durability.

11.8.4   Stainless Steel Tank

These tanks are made of stainless steel sheets with thickness varying from 0.6 to 3 mm depending on the size ranging from 200 l [18] to >1 million litres [19] as shown in Fig. 11.23. These tanks are durable, highly resistant to corrosion, and almost maintenance-free, however, they are costlier.

Fig. 11.23 Stainless steel water tank

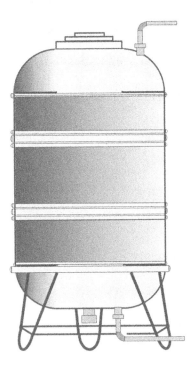

Fig. 11.24 Plastic water tank

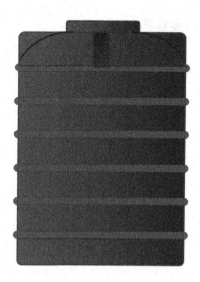

11.8.5   Plastic Tank

These tanks are made of fiberglass-reinforced plastics, high-density polyethylene, or other plastic materials. The tanks are relatively light, easy to carry, and easy to install; however, they are not durable due to the deterioration of plastic quality from the Sun's rays. A plastic tank of maximum size thus far manufactured has a 45,000-l capacity [20]. All plastic tanks used for storing rainwater to be used for culinary or drinking purpose must be made of food-grade plastic. A vertical-type plastic tank, as shown in Fig. 11.24, should be placed on a masonry base, which must be constructed in such a way that it supports the entire bottom surface of the tank.

11.9   Pump

In general, floor-mounted centrifugal pumps are used for rainwater supply in buildings. However, in limited cases a submersible pump can be used. The use of a submersible pump poses some advantages and disadvantages as well.

11.9.1   Submersible Pumps

Submersible pumps are basically a centrifugal-type pump mostly designed for vertical installation. Therefore, these pumps are installed vertical keeping them fully submerged in the water to be pumped. In a submersible pump, the motor and pump

Fig. 11.25 Submersible
pump

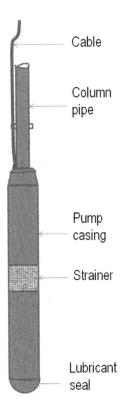

Cable

Column
pipe

Pump
casing

Strainer

Lubricant
seal

units are assembled in a single unit as shown in Fig. 11.25. The electric motor remains protected by a water-proof housing. Bearings are usually lubricated by water or water–oil emulsion, and leakage is protected by mechanical seal. Power to the motor is fed by a waterproof cable clipped to the side of the rising main.

Advantages of a submersible pump:

1. Exceptionally quiet while running
2. No priming required
3. No space required for housing the pump
4. Quick and easy installation

Disadvantages of a submersible pump:

1. Comparatively less efficient
2. Maintenance is troublesome.

11.9.2 Pump Installation

Proper installation of a pump is one of the most important jobs to be performed for its better performance. The following important factors should be considered when installing a pump.

Location: The pump should be located near to the source of rainwater where it is accessible, and sufficient light should be available for inspection of the pump particularly the condition of the packing and bearings for leakage.

Foundation: The Foundation should be rugged enough to afford permanent rigid support to the base area of the bed plate and to absorb all stress and strains. For this purpose, a mass-reinforced concrete foundation is most satisfactory. The general rule is that the weight of the foundation should be at least equal to the weight of the pump set. Foundation bolts should be correctly positioned according to the drawing supplied by the manufacturer.

Grouting: Grouting, 20–50 mm-thick with a 1:2 cement-to-sand ratio, should be poured under the bed plate to provide a solid bearing for the pump. The grouting should be allowed to set for at least 48 h.

Leveling: The pump and motor should be accurately aligned to minimize vibration and load on the pump. Vertical pumps should be truly vertical, and horizontal pumps should be truly horizontal. Alignment is checked by checking the faces of the coupling halves of the pump and the motor for parallelism with the help of a feeler gauge and straightness of the sides and leveling of tops using a straight edge.

Suction-pipe requirements: Suction pipe should be as short and direct as possible. A suction pipe larger than the suction inlet of pump should be connected with a pump inlet by an eccentric reducer and a flexible connector. The pipe should be slightly graded downward, and, if needed, bending can be done at a distance of 12 times the diameter of the suction pipe apart from the suction inlet of pump as shown in Fig. 11.26. A strainer having an opening area 3 times the suction pipe cross-sectional area should be installed at the end. All joints should be tested for air tightness.

Delivery-pipe requirements: A delivery pipe should have a minimum number of bends. The delivery pipe should also be fitted with a pump outlet by a flexible connector. A gate valve and a check valve, when the riser pipe is >10 m, should be installed near the pump (Fig. 11.27). A pressure gauge is also installed on the delivery pipe to monitor the pressure in the system.

11.9.3 Pump Room

A pump room should be sufficiently spaced to accommodate the required number of pumps along with its piping and other accessories. In addition, sufficient working space is also needed for easy operation and maintenance. A minimum 1.2-m

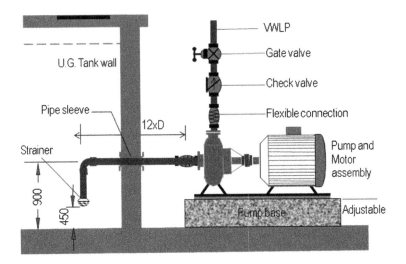

Fig. 11.26 Various elements in pump installation

Fig. 11.27 A gate valve

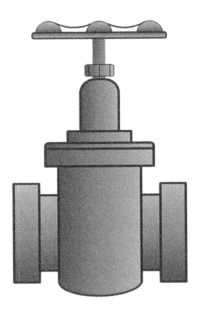

clearance should be provided near the foundation, or between foundations, as a working space. For pumps >40 BHP, sufficient overhead arrangements for lifting and lowering the pump should be provided. The area of a pump room should be provided considering the number and sizes of pumps to be housed in as listed in Table 11.2.

Table 11.2 Space requirement for electric-driven pumping rooms

Number and brake horse power		1–10	2–10	1–25	2–25	1–40	2–40	1–100	2–100	2–100 fire pump
Space (m²)	Horizontal shaft	6	8	9	14	13	20	25	40	90
	Vertical shaft	6	8	7	12	10	14	12	16	–

Source [21]

The pump room should be sufficiently ventilated and illuminated. A sufficient and proper drainage system should be provided for the wastage or leakage water. The pump room should be constructed where there is no danger of flooding and should be away from occupancies of low sound tolerance.

11.10 Valve

Valves are devices used to control the flow of fluid mostly in the piping. Virtually a valve is a device having a lid or cover to an aperture such that it can be operated manually or automatically, e.g., opened fully or partially to regulate, direct, or control the flow of fluid (liquids, gases, fluidized solids, or slurries) by opening, closing, or partially obstructing the passageways through lifting, turning, or sliding the lid or cover to attain a particular condition of flow. In addition to valves, many other appurtenances exist such as faucets and cocks, etc., which are mostly appertained to the piping of rainwater-supply system.

Valves, mostly used in rainwater supply in buildings, may be classified into two broad categories.

1. Non-automatic and
2. Automatic.

Nonautomatic valves are operated manually. The commonly used valves in this group are as follows.

1. On–off valve
2. Throttling valve and
3. Flushing valve.

Automatic valves are controlled by some condition of services operating to open or close those. The commonly used valves in this group are as below.

1. Check valve
2. Float valve
3. Pressure-reducing valve and
4. Air release valve.

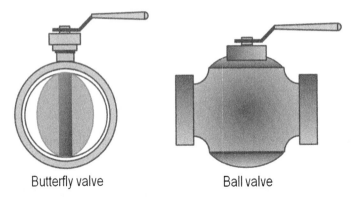

Butterfly valve   Ball valve

Fig. 11.28 Various on–off valves

11.10.1 On–off Valve

With the help of this type of valve, flow is controlled by a shut-off screw spindle to move a gate to the open and closed position. Therefore, on–off valves are commonly called "gate valves".

Various common types of on–off valves, shown in Fig. 11.28, are as follows.

1. Gate valve
2. Butterfly valve
3. Ball valve, etc.

11.10.2 Throttling Valve

A throttling valve is primarily used to regulate, or throttle, the flow of fluid in the pipe and secondarily to perform as on–off valve. It is also called a "throttling valve". Flow in the valve occurs in tortuous path. Therefore, the friction loss is comparatively high.

A common flow-throttling valve is shown in Fig. 11.29. Various flow-throttling valves are as follows.

1. Globe valve and
2. Angle globe valve.

Fig. 11.29 Throttling valve

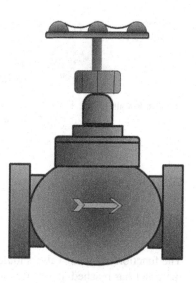

Fig. 11.30 Check valve

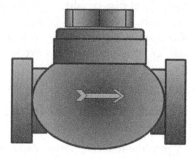

11.10.3 Check Valve

A check valve is installed to permit the flow of fluid within the pipe in one direction and to close automatically when the flow is in the reverse direction. A check valve is shown in Fig. 11.30. Various check valves used are as follows.

1. Swing-check or flap valve and
2. Lift-check valve.

Swing-check valve causes very little friction loss, whereas a lift-check valve causes considerably high friction loss.

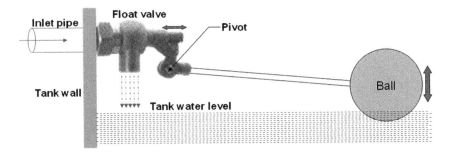

Fig. 11.31 Float valve

11.10.4 Float Valves

The function of a float valve should shut off the water supply to a water tank when the water has reached a predetermined level. This valve consists of a ball as shown in Fig. 11.31, which floats on the water surface and changes its position as the water level rises or falls. When the water level starts falling due to consumption, the ball starts going down, thereby starting water inflow through the tank inlet pipe. Conversely, when water is poured into the tank, the water level rises and lifts the ball up, thus pushing the valve to shut off the inflow of water into the tank.

11.10.5 Pressure-Reducing Valve

A pressure-reducing valve, shown in Fig. 11.32, is installed to maintain a desired (preset) constant pressure in the downstream of the valve. The pressure-reducing valve has an adjustment screw at the top of the valve to adjust the outlet pressure. Pressure reducing from the upstream side of the valve to the downstream side can have a ratio as high as 10:1. However, an approximately 25 % pressure reduction is preferable.

11.10.6 Air-Release Valve

An air-release valve, shown in Fig. 11.33, is used in the apex points of a water supply-piping system for releasing trapped air from the piping, which can restrict flow and thereby increase pumping costs, thereby decreasing the efficiency of the rainwater-supply system. The size of the air-release valve should be approximately one fourth of the size of the pipe on which it is installed.

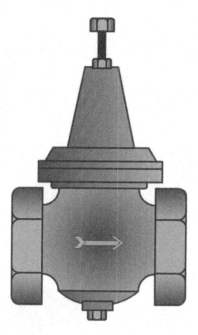

Fig. 11.32 Pressure-reducing valve

11.10.7 Flush Valve

A flush valve, also called a "flushometer," (Fig. 11.34), is a device designed to supply a predetermined quantity of water directly from the water-supply pipe for flushing the water in a water closet or urinal bowl. With this device, flushing can be done at 6- to 10-s intervals. Therefore, these valves are recommended for public toilets. Generally, two sizes of flush valve are used: One is a 25-mm size used for flushing in a water closet and other is a 19-mm size used for flushing urinals. The 25-mm flush valve discharges approximately 4.5 l of water/flush, whereas a 19-mm flush valve discharges approximately 3.75 l of water/flush.

11.11 Cock

A cock is a type of valve intended to form a convenient means of shutting off the flow of water out of a pipe. It is also called a "stop cock". Cocks are classified as below and are shown in Fig. 11.35.

1. Straight-way cock
2. Angle stop cock and
3. Waste or drain cock.

Fig. 11.33 Air-release valve

Fig. 11.34 Flush valve

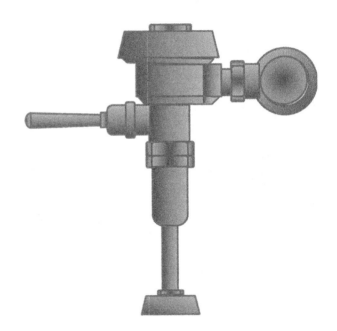

Angle stop cock   Straight way cock   Drain cock

Fig. 11.35 Stop cocks

11.12 Faucet

A faucet and bib are generally used to signify a valve controlling the outlet of a pipe that conveys water. Faucets are used generally for indoor purposes. In buildings where rainwater is harvested, obviously water-efficient and less water-consuming faucets should be used.

With respect to the mode of operation, faucets are classified as below.

1. Compression faucets

   (a) Plain faucet and
   (b) Self-closing faucet

2. Fuller faucet and
3. Mixture faucet.

11.12.1 Compression Faucets

In this type of faucet a disc type washer is compressed against a seat by a stem operated by a handle to close the flow of water. Therefore, these faucets are grouped under compression faucets. Compression faucets are of two categories as follows.

Plain faucet: In this type of faucet, a disc washer is used, which is attached at the end of a stem maneuvered by a handle fitted on its top. There is a solid or removable seat inside. When the handle is turned off, the washer at the end of the stem rubs and compresses against the seat to close off the water flow. By turning the handle counter-clockwise, the washer is lifted up to allow the flow.

Plain faucets are also called a "tap" or "cock". There are two types of plain faucets, Bibb cock and Pillar cock, as shown in Fig. 11.36. Bib cocks are designed to be installed on vertical surfaces and pillar cocks on horizontal surfaces.

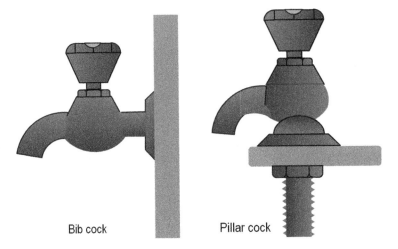

Bib cock   Pillar cock

Fig. 11.36 Plain faucets

Fig. 11.37 Self-closing
faucet

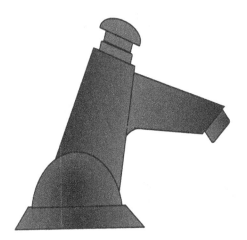

Self-closing faucet: Self-closing faucets are spring-loaded faucets that close automatically. Figure 11.37 shows a typical self-closing faucet. The handle should be pushed or operated to obtain the water flow. When the handle is pushed, water starts flowing; after some time, the flow will be automatically stopped. Self-closing faucets of varying running periods—e.g., approximately 7, 15, or 30 s after every press of a button—are available. These types of faucets are used to minimize the consumption of water by preventing wastage. By using these types of faucets, approximately 30 % water can be saved compared with water consumption using plain faucets.

Fig. 11.38 Fuller faucet

11.12.2 Fuller Faucet

The Fuller faucet, as shown in Fig. 11.38, is a very desirable type of faucets for use on low-pressure lines. By turning the handle by 90°, the faucet can be fully opened quickly. Therefore, in a high-pressure line, the use of such a faucet may cause "water hammer," which will create sound and vibration in the pipe.

11.12.3 Mixture Faucets

Mixture-faucets allow the mixing of cold and hot water to the individuals' preference. Instead of two separate faucets, one for cold and one for hot water flow, the mixture faucet combines cold- and hot-water valves with a single spigot to supply water of the desired temperature through a single spout. In a manually controlled mixture faucet, the temperature is controlled by manipulating each valve with separate handle or a single lever by turning it aside. Such faucets are usually satisfactory on a basin, a bathtub, a sink, etc. Figure 11.39 illustrates a single-lever mixture faucet. In mixture faucets, the pressure at the inlet of the cold and hot water should be maintained more or less same.

11.13 Water Meter

A water meter is a device that measures the volume of water that passes through a water-service pipe. It may be required to know the quantity of rainwater harvested in a building, for which the installation of a water meter might be necessary. Three types of water meter are available.

1. Disc meter
2. Turbine meter and
3. Compound meter.

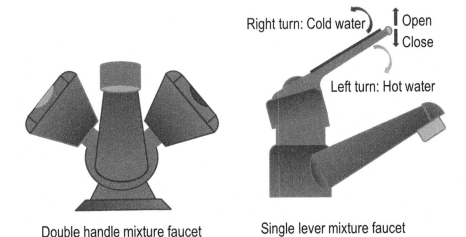

Double handle mixture faucet   Single lever mixture faucet

Fig. 11.39   Mixture faucets

11.13.1   Disc Meter

Disc meters, as shown in Fig. 11.40, are mostly used in a small amount of water flow in pipes with diameter ≤50 mm. These meters give an accurate measurement of flow.

11.13.2   Turbine Meter

Turbine meters are used in a water-service pipe in which a large and constant volume of water flows. In this type of meter, the velocity of flow is considered in measuring the volume of water, so it is also called a "velocity meter".

Fig. 11.40   Disc meter

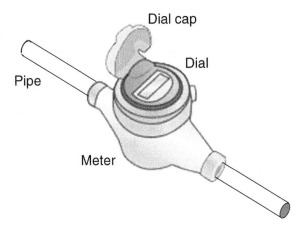

11.13.3   Compound Meter

A compound meter is a combination of a disc meter and turbine meter. It is used where high fluctuation occurs in a large volume of water flow. The size of this type of meter varies from 50 to 250 mm.

11.14   Elements in Recharge Structures

Recharge structures, such as recharge pits, troughs, wells, etc., are constructed of brick or cement blocks using a mortar made of cement and sand as binding material. On the top of these structures, an RCC slab is provided to cover the structure. In the base of the pits, a CC or RCC slab is provided. RCC elements use cement, aggregates, and reinforcing bars. In all masonry works, water is a vital ingredient. In recharge structures, one or more manhole covers are placed to inspect or enter the structure, where necessary, through one of the holes.

11.14.1   Bricks and Cement Blocks

Bricks are hard blocks, as shown in Fig. 11.41, made by burning selected clay; they are mostly used as building material in many countries. Bricks are also used in rainwater-recharging structures. Depending on the quality and configuration of shape, bricks are classed as first, second, or third class. In making groundwater-recharging structures, first-class bricks should be used; however, for economy second-class bricks may be also used. Specifications of clay bricks to be used are as follows.

1. Size: 62 mm × 238 mm × 112 mm (thickness)
2. Weight: 3.2 kg
3. Crushing strength: 21 MPa
4. Water absorption: 12–20 %

Cement blocks can also be used as an alternative to bricks in masonry construction for rainwater harvesting. Cement blocks are machine-made using cement and sand as the basic material at a ratio of 1:7–1:9, respectively. These blocks may

Fig. 11.41 Clay brick

Fig. 11.42 Cement blocks

be solid; however, they are mostly manufactured with one or more hollow cavities as shown in Fig. 11.42. Concrete blocks are held together with cement mortar to form the desired wall. Deep and large recharge structures should be constructed using solid blocks; hollow blocks may be used in small recharge and inspection pits.

Specification of hollow cement blocks:

1. Size: Height = 150, 200, or 250 mm; width = 200 mm; and length = 400 mm
2. Weight: 17.2–19.5 kg
3. Compressive strengths: 2.9–22.5 N/mm^2

Specification of solid cement blocks:

1. Height = 50, 75, or 100 mm; width = 200 mm; and length = 400 mm.
2. Dry density: 2000 kg/m^3.
3. Compressive strength: 2.9–40 N/mm^2.

11.14.2   Cement

Cement is a dry powder-like material containing predominately tri-calcium silicate. Cement consists of mainly lime 75–77 %, silica and alumina 12–15 %, and iron oxide 0.5–6 %. *Ordinary Portland cement (OPC)* has been the most widely used cement since its appearance in 1824. Now cements containing pozzolonic materials, such as fly ash, slag, volcanic ashes, etc., are available and commonly known as Pozzolanic Portland cement (PCC). Because it has a low percentage of pozzolanic materials, PCC may also be used in building recharge structures. Cement is normally available in 50-kg bags.

Mortar or concrete, and whatever else is made of cement, gains its strength through a chemical process with water called "hydration." Hydration is a complex exothermic-reaction process between water and the cement ingredients, from which the final product, i.e., concrete, derives great strength.

11.14.3   Aggregate

Aggregates—which are a mix of inert granular materials such as sand, gravel, or crushed stone of varied sizes—are used in making concrete. Crushed clay brick chips are also used as an aggregate.

Generally approximately 60–75 % of the total volume of cement concrete is comprised of aggregates. Aggregates in concrete are divided into two distinct categories: fine and coarse aggregates. Fine aggregates generally constitute natural sand or finely crushed stone having particles mostly passing through a 9.5-mm sieve. Coarse aggregates are any particles >4.83 mm; however, they generally range between 9.5 and 37.5 mm in diameter. For making durable concrete, aggregates must be clean, hard, strong, rough surfaced, and free of chemicals. In addition, aggregates must be free of clay and other foreign materials that can cause the deterioration of concrete.

11.14.4   Sand

Sand is a naturally occurring granular inert material composed of fine particles of rocks and minerals. Sand is used as a fine aggregate in making mortar and concrete. In terms of particle size, as used by geologists, sand particles range in diameter from 0.0625 (or 1/16 mm) to 4 mm. The other way of considering granular matter, such as sand, is those particles that pass through a no. 200 (0.075-mm) sieve and retain on a no. 4 (4.75-mm) sieve. Sand used in construction should be free of dirt and other foreign materials.

11.14.5   Reinforcing Bar

Mild-steel bars of various sizes are used for the reinforcement of masonry structural elements, mainly concrete, to increase its flexural strength. They are produced by a hot-rolling process with subsequent superficial hardening by heat treatment.

Concrete is weak in tension but strong in compression. Therefore, reinforcing steel bars are used to increase the tensile strength of concrete. Considering the surface finish, reinforcing bars (rebars) are produced either plain or deformed as shown in Fig. 11.43. Deformed steel bars are normally designed with several ridges, in regular pattern, that help it to be more firmly anchored within concrete or masonry. Currently deformed bars are mostly used. Reinforcing bars are available in sizes varying from 10- to 50-mm diameter. In recharging structures, depending on the loading condition, the sizes of bar that might be used are mostly 10–16 mm in diameter.

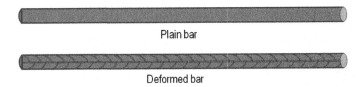

Fig. 11.43 Plain and deformed reinforcing bars

Reinforcing steels are manufactured in different grades yielding different degrees of strength. Yield strength is the tensile stress at which a predetermined amount of permanent deformation occurs in the steel section when stretched. Reinforcing steel, specified as 40 grade, has a yield strength of 275 mPa; 60-grade steel has a yield strength of 420 mPa. In recharging structures, 40-grade steel is sufficient.

11.14.6 Cement–Sand Mortar

Cement–sand mortar is a workable paste used to bind masonry blocks together and fill the gaps between them. Such mortar is prepared by mixing cement and sand in different proportions with water in definite proportions. A variation in the mix proportion of cement and sand causes a variation in the crushing strength of mortar. Generally increasing the amount of cement increases the strength of the mortar. Beyond a certain point, it also acts negatively by making the mortar very brittle. Because cement is expensive, a lesser amount of cement should be used to optimize cost in low-strength applications. Again a very high amount of sand will make the mortar very brittle and weak against all kinds of stress.

Generally the mixed proportion of cement and sand is a ratio of 1:6. Mortar becomes hard when the mix sets, thus resulting in a rigid aggregate structure. After adding water to the dry mix of cement and sand, the mixture should be well blended and used within half an hour of mixing. A minimum ratio of water to cement (w/c) of approximately 0.25 by weight is necessary to ensure that all the cement particles come into contact with the water to complete the hydration process. The addition of excess water will result in poor-quality mortar.

11.14.7 Concrete

Concrete is a composite material composed of inert granular materials embedded in a hard matrix of material and cement, as binder, which glues them together by chemically reacting with water. Generally cement, aggregates, and water are the essential ingredients in making concrete. Cement concrete should be used under small pits or chambers. In making plain cement concrete for these purposes, a 1:2:4 ratio, respectively, of cement, sand, and coarse aggregate is appropriate. This mix proportion will comprise 1 part cement, 2 parts sand, and 4 parts coarse aggregate by volume. The key to achieving strong and durable concrete depends on the careful proportioning and mixing of the ingredients with the optimum quantity of water. Through a chemical reaction of cement with water, the cementation paste hardens and gains strength with time ultimately to form the rock-like hard concrete.

11.14.8   Water

The principal role of water in making mortar and concrete using cement has already been discussed. Water is also used for various other purposes in masonry and concrete works as mentioned below.

1. Cleaning aggregates if dirt particles are present
2. Soaking clay bricks and aggregates before use
3. Improving workability and
4. Curing all products made of cement.

Water to be used in masonry or concrete work should be clean and cool. It should also be free of elements such as oils, acids, alkalis, salts, or any organic materials that are injurious to concrete. Virtually potable water is generally considered satisfactory for any cement product. The pH value of the water should not be <6.

Water plays various others role in achieving some good properties of cement concrete. It confers increased strength, resistance to weathering, and a good bond between concrete and reinforcement. Water also plays role in decreasing shrinkage and cracking due to quick drying, changes in volume due to wetting and drying, etc.

By adding water in cement concrete workability is improved, which helps in transporting long away, placing and finishing without segregation easily and comfortably. The addition of excess water causes bleeding in concrete, which increases its permeability.

Considering these factors, the ratio of water to cement should be strictly monitored and maintained. A typical range of ratios for water to cement (w/c) is 0.25–0.6. In making mortar and concrete for recharging and drainage structures, a w/c ratio of 0.45 may be desirable.

Ideally all exposed surface of newly made mortar and concrete should be kept wet or damp for a long period until the hydration process is complete. The process of keeping an exposed surface of newly made mortar and concrete wet or damp is termed "curing", which is very important in gaining strength with time and achieving good quality. Curing should be carried out for a period of time not less than 14 days from the date of mixing and pouring and should not last >28 days. Curing is done in various ways as described below.

Fig. 11.44 Reinforcement of concrete slab over pit

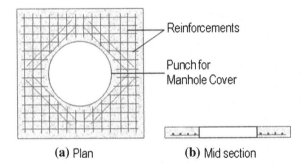

Reinforcements

Punch for
Manhole Cover

(a) Plan   **(b)** Mid section

1. Keep the mortar or concrete submerged under water.
2. Spray water on the concrete or mortar surface at regular intervals.
3. Cover the mortar or concrete with a layer of wet sacks, canvas, hessian, or similar materials, which should be kept constantly wet during the curing period.

11.14.9   Reinforced Cement Concrete

Concrete has a relatively high compressive strength; however, it has much lower tensile strength. For this reason it is usually reinforced with mild-steel bars, which are strong in tension. Concrete has a very low coefficient of thermal expansion, and it shrinks as it matures. All concrete structures are susceptible to cracking to some extent due to shrinkage and tension. To avoid or minimize these effects, reinforcing steel plays the key role.

Generally groundwater-recharging and -drainage structures are covered with reinforced concrete slabs. There is a chance of having various live loads on the structure, in addition to its own weight, and thus tension will increase at the bottom portion of the slab. Therefore reinforcing bars should be provided as shown in Fig. 11.44. The minimum reinforcement required to withstand the stress of contraction and expansion due to thermal changes is 0.25 % of the concrete section. Considering other load factors, the reinforcement ratio can be considered as 0.5–0.75 % of the concrete section.

11.15   Filter Media

A filtering medium, which consists of sand and gravel, is used in filtering rainwater for general purposes of use and for recharging groundwater. The sand layer is placed over the gravel layer. The quality and specification of these materials are discussed below.

11.15.1 Filtering Sand

In filtering rainwater, sand is an extremely effective filter medium because of its ability to hold back precipitates containing impurities. Uniformity in size, angularity, and hardness are the important characteristics of filtering sands to ensure efficient filtering. This sand should be free of clay, loam, vegetable matter, organic impurities, etc., and should be gap-graded, i.e., uniform in nature and grain size. Instead of sand, "anthrafilt", which is made from anthracite (stone-coal), can also be used as filter medium that offers a high rate of filtration, better efficiency, and low cost. The physical specifications of filtering sand should be as follows [22]:

1. Effective size: 0.20–2.5 mm
2. Uniformity coefficient: 1.3–1.7
3. Specific gravity: 2.67
4. Moh's hardness: 7
5. Density: 1681.94–1842.12 kg/m^3.

11.15.2 Filtering Gravel

Gravel is granular soil material which passes through a no. 4 (4.74-mm) sieve. Filtering gravel is also an effective filter media because of its ability to hold back precipitates containing impurities. The gravel used should be clean and free of clay, dust, silt, and organic matter, and it should be durable, hard, round, and strong. It is usually graded and laid in layers of 150 mm. The size of the gravel increases from top to bottom. The physical specifications of filtering gravel should be as follows [23]:

1. Effective size: 3–40 mm
2. Uniformity coefficient: 1.2–1.7
3. Specific gravity: 2.70
4. Moh's hardness: 6
5. Density: 1601.85 kg/m^3.

11.16 Drainage Structures

Two elements are incorporated into a rainwater- or stormwater-drainage system.

1. Drainage pipe and
2. Manhole or inspection chambers.

11.16.1 Drainage Pipe

Stormwater-drainage pipes in common use fall into two groups as follows:

1. Flexible pipe, which includes corrugated metal and plastic pipe
2. Rigid pipe, which includes reinforced-concrete pipe, plain-concrete pipe, clay pipe, cast-iron pipe, etc.

ABS plastic pipe should be used as a drainage pipe. Use of this plastic pipe is preferable where there is a chance of soil subsidence under the pipe. In case of a large-diameter drainage pipe, plastic pipe may not be cost-effective.

RCC pipe is a rigid drainage pipe that has significant inherent strength to support loads without aid from the backfill. However, its load-carrying capacity can be increased by taking special measures in bedding, side filling, and backfilling the pipe.

Laying of pipes: Underground pipes should be laid in a very straight line to a steady gradient as designed. A taut string line, sight rails, or a laser line is generally used to ensure true alignment and the grade of pipe. The trench should be excavated in advance, and the bed should be prepared with a scooped-out recess to accommodate sockets or collars and the shape of the pipe to provide even bedding of the pipeline. The width of the trench should be cut adequate to allow room for a working area and for tamping the side support. It should be not <200 mm wider (100 mm each side) than the outside diameter of the pipe irrespective of soil condition. For larger pipes, the trench bottom, supporting the pipe, should be made curved matching the curvature of the pipe as far as possible. The arc length of curvature of the trench bottom, remaining in contact with the pipe, should subtend an angle of approximately 120° at the centre of the pipe as shown in Fig. 11.45. The depth of trenches should be excavated considering the specified depth of the bedding, the pipe diameter, and the minimum recommended cover constituted by overlay plus backfill above the pipe.

Pipes should be laid on a full bed of granular material, either gravel or sand, containing little or no fines. The absolute minimum underlay should be 75–100 mm. The pipe-overlay material should be leveled and tamped in layers to a minimum height of 150 mm above the crown of the pipe. The pipe side support should also be adequately tamped in layers ≤150 mm unless otherwise specified. The pipe side-support material, pipe-overlay material, and pipe-bedding material used should be identical in nature and quality.

Sand, which is free of clay, organic matters, sharp objects, etc., as well as gravel of approved grading up to a maximum size of 14 mm are suitable as pipe-bedding material. Unless otherwise specified, excavated material from the trench can be used as backfill if it is easily compactable. Before using excavated material as backfill, it should be made free of any injurious foreign materials. The recommended minimum cover of backfill under non-vehicular loading is 300 mm. For under-vehicular loading, the backfill-cover depth will vary from 450 to 750 mm depending on the expected vehicular load [24].

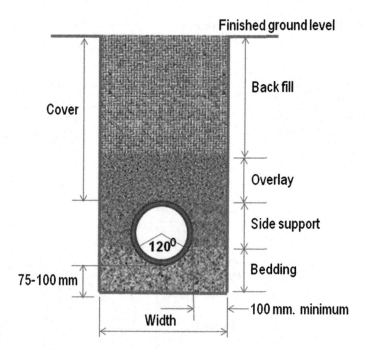

Fig. 11.45 Dimensions of trench and fills for laying underground rigid pipe

Ditches, pot holes, etc., must be carefully backfilled to regain the stability of the filled soil. Care should be taken that the voids or zones of poor compaction are not left under the haunches inside the bell holes of pipe.

11.16.2   Inspection Pit or Manhole Construction

Inspection pits or manholes are constructed at locations where needed. Manholes or inspection chambers made of bricks or cement blocks have side walls typically 200- to 250-mm thick. To make smaller and strong cover, the top dimension of the chamber is gradually decreased by converging the wall as shown in Fig. 11.46. Chambers and manholes are constructed on a bed slab of mass concrete or RCC at least 150-mm thick. A precast or cast-in-place RCC slab with a manhole cover is placed over the chamber. The bottom-most outlet pipe should be placed at least 75–150 mm above the bottom surface of the pit.

When installing pipes in manhole walls, usually a heavy-duty percussion drill is used to remove bricks or blocks at that locations; alternatively, "stitch-drilling" can be used for this purpose. In these chambers, it is preferred to make the bottom parts of the pits first and then the lower portion of the brick or block works are

Fig. 11.46 Section of a
manhole or pit

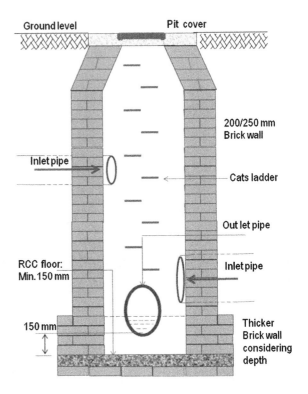

constructed. The pipes to be inserted are then placed on the walls at the desired invert levels. Later the brick or block works in between the pipes and above the pipes are completed.

Precast concrete circular manhole sections are usually made of a 50–75 mm-thick wall. It is recommended to make "stitch-drill" for these sections to prevent fracture or spoiling of individual sections during the creation of holes for connecting pipes in the manhole. A 13-mm masonry bit should be used to drill holes in the circumference of the circle that have a larger diameter than the pipe to be connected. By hammering and chiseling, the concrete within the stitch-drilled boundary can then be broken out easily with no danger and making no catastrophic damage in the chamber wall.

11.16.3 Manhole Cover

Manhole covers are the lids installed on the cover slab placed on the inspection chambers or manholes. These covers are available in a wide variety of shapes, sizes and materials. The lids may be solid or perforated. The openings in the lid, as shown in Fig. 11.47, should allow stormwater to enter the chamber or manhole.

Fig. 11.47 Manhole cover
for storm drainage

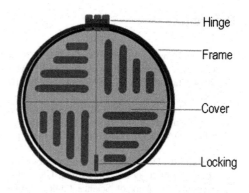

Hinge

Frame

Cover

Locking

Special attention should be paid to the covers when the manholes are located in paved areas. Manhole covers mostly consist of two components: the cover itself and a frame inside which the cover fits. The frame is installed to a specified height set to suit the surrounding paved or ground level. The cover is hinged with the frame so it can be opened when access into the manhole is required.

For inspection of the chamber, the manhole cover should not be <450 mm. For special purposes, a manhole cover with a smaller diameter may be used. The minimum diameter of the manhole cover should be 600 mm. Manhole covers should be sufficiently strong to withstand test load ranging between 15 and 900 kN [25], and it should be long-lasting, especially in a corrosive environment. Heavy-duty manhole covers should be used subject to the anticipated traffic or heavy load passing over it. Otherwise light-duty covers can be used in all pedestrian and non-pedestrian areas. Cast-iron manhole covers are widely used. Various classes of manhole covers with respect to load test are listed in Table 11.3.

Table 11.3   Various classes of manhole covers and their test loads according to BS EN 124:1994

Group	Class	Load bearing capacity (kN)	Permitted area types
Group 1	A15	15	In areas where only pedestrians have access
Group 2	B125	125	In car parks and pedestrian areas where only occasional vehicular access is likely
Group 3	C250	250	In areas with slow-moving traffic; in highway locations up to 500 mm from the curb and up to 200 mm into the verge excluding motorways
Group 4	D400	400	In areas where cars and lorries have access including carriageways, hard shoulders, and pedestrian areas
Group 5	E600	600	In areas where high wheel loads are imposed, e.g., loading areas, docks, aircraft pavements, etc.
Group 6	F900	900	In areas where particularly high wheel loads are imposed such as aircraft pavements

Source [25]

References

1. National Capital Region Planning Board (NCRPB) Characteristics of various plastic, DI and GI pipes. http://ncrpb.nic.in/NCRBP%20ADB-TA%207055/Toolkit-Resources/Annexure%2014_Characteristics%20of%20plastic%20DI%20&%20GI%20pipes.pdf. Retrieved on 28 Oct 2015
2. IPS Flow Systems (2015) About:pvc-u. http://www.ipsflowsystems.com/pdfs/pvcu/pvcuabout.pdf. Retrieved on 29 Oct 2015
3. Chasis DA (2014) Plastics and sustainable piping system. Industrial Press Inc., New York, p 223
4. Haugh MC, Dolbey R (1995) The plastics compendium, vol 1. In: Key properties and sources. Rapra Technology Ltd. Shropshire, UK, p 3
5. National Academy of Science (2006) Drinking water distribution systems: assessing and reducing risks. The National Academies Press, Washington D.C., USA, p 37
6. Bureau of Indian Standard (BIS) (2001) Centrifugal cast (Spun) iron pressure pipes for water, gas and sewage. In: Specification, 4th revision. https://archive.org/stream/gov.in.is.1536.2001/is.1536.2001_djvu.txt. Retrieved on 02 Feb 2016
7. Ontario Concrete Pipe Association (OCPA) (2015) Concrete pipe design manual. http://www.ocpa.com/_resources/OCPA_DesignManual.pdf. Retrieved on 30 Oct 2015
8. Cemex Company (2015) Life of concrete pipe. Info brief. http://www.rinkerpipe.com/TechnicalInfo/files/InfoBriefs/IB1003LifeConcretePipe.pdf. Retrieved on 30 Oct 2015
9. Kansas Department of Health and Environment (2008) Appendix C: procedure for pressure and leakage testing of water mains. http://www.kdheks.gov/pws/mds/Appendix_C.pdf. Retrieved on 30 Oct 2015
10. Housing and Building Research Institute (HBRI) (1993) Bangladesh National Building Code (BNBC)
11. Harris CM (1991) Practical plumbing engineering. McGraw-Hill Company, New York
12. Fallbrook Public Utility District (FPUD) (2012) Disinfection of pipe and water storage facilities, Section 15041, FPUD Standard specifications. https://www.fpud.com/PDFDocuments/Construction%20Standards/Section%203%20Technical%20Specifications/15041%20-%20Disinfection%20of%20Pipe%20&%20Water%20Storage%20Facilites.pdf. Retrieved on 30 Oct 2015
13. Metkar SM (2015) Economics of R.C.C. water tank resting over firm ground vis-a-vis pre-stressed concrete water tank resting over firm ground. Civil engineering forum. http://www.engineeringcivil.com/economics-of-r-c-c-water-tank-resting-over-firm-ground-vis-a-vis-prestessed-concrete-water-tank-resting-over-firm-ground.html. Retrieved on 30 Oct 2015
14. Ammas (2015) Water storage tanks. Response of Preeti Saxena. http://www.ammas.com/q&a/Water-storage-tanks/q/185752. Retrieved on 1 Nov 2015
15. Housing and Building Research Institute (HBRI), Flyer on products of HBRI
16. Sharma PC (2015) Ferrocement water storage tanks for rain water harvesting in hills and islands. http://www.eng.warwick.ac.uk/ircsa/pdf/12th/6/PCSharma.pdf. Retrieved on 06 Nov 2015
17. Rain Harvesting Supplies, Inc (2015) Metal water tanks. https://www.rainharvestingsupplies.com/metal-water-tanks/. Retrieved on 1 Nov 2015
18. Osmonic (2015) Stainless steel (SS) water and liquid storage tank manufacturer. http://www.osmonicwater.com/stainless_steel_water_tank.html. Retrieved on 1 Nov 2015
19. Bestank (2015) Stainless steel water storage tanks. http://bestank.com/products/water-tanks/. Retrieved on 1 Nov 2015
20. Portable Tanks (2015) Large water storage tanks. http://www.water-storage-tank.com/large-water-tanks.html. Retrieved on 1 Nov 2015
21. Jain VK (1985) Handbook of designing and installation of services in high rise building complexes. Jain Book Agency, India, p 255

22. Kleen Industrial Services (2015) Improving water quality globally. http://www.kleenindustrialservices.com/water-filtration-media-silica-sand.html. Retrieved on 2 Nov 2015
23. Kleen Industrial Services (2015) Improving water quality globally. http://www.kleenindustrialservices.com/water-filtration-media-filter-gravel.html. Retrieved on 2 Nov 2015
24. Vinidex Pty. Ltd (2013) Below ground installation. http://www.vinidex.com.au/technical/pvc-pressure-pipe/below-ground-installation/. Retrieved on 6 Nov 2015
25. Wrekin Products (2015) Covers and frames. http://www.wrekinproducts.com/british-standard-groups-and-classes/. Retrieved on 2 Nov 2015

Appendix

See Table A.1, A.2, A.3, A.4, A.5, A.6, A.7, A.8, A.9, A.10, A.11, A.12, A.13, A.14, A.15, A.16, A.17, A.18 and A.19

Table A.1 Country wise average precipitation in depth (mm per year) A-C

Country name	2011	Country name	2011	Country name	2011
A		**B**		**C**	
Afghanistan	327	Bahamas, The	1292	Cabo Verde	228
Albania	1485	Bahrain	83	Cambodia	1904
Algeria	89	Bangladesh	2666	Cameroon	1604
American Samoa		Barbados	1422	Chile	1522
Andorra		Belarus	618	China	645
Angola	1010	Belgium	847	Cayman Islands	
Antigua and Barbuda	1030	Belize	1705	Central African Republic	1343
Argentina	591	Benin	1039	Chad	322
Armenia	562	Bermuda		Colombia	2612
Aruba		Bhutan	2200	Comoros	900
Australia	534	Bolivia	1146	Congo, Dem. Rep	1543
Austria	1110	Bosni and Herzegovina	1028	Congo, Rep	1646
Azerbaijan	447	Botswana	416	Costa Rica	2926
		Brazil	1782	Cote d'Ivoire	1348
		Brunei Darussalam	2722	Croatia	1113
		Bulgaria	608	Cuba	1335
		Burkina Faso	748	Curacao	
		Burundi	1274	Cyprus	498
				Czech Republic	677

© Springer International Publishing Switzerland 2017
S.A. Haq, PEng, *Harvesting Rainwater from Buildings*,
DOI 10.1007/978-3-319-46362-9

Country wise average precipitation in depth (mm per year) D-K

Country name	2011	Country name	2011	Country name	2011
D		**G**		**I**	
Denmark	703	Gabon	1831	Iceland	1940
Djibouti	220	Gambia, The	836	India	1083
Dominica	2083	Georgia	1026	Indonesia	2702
Dominican Republic	1410	Germany	700	Iran, Islamic Re.	228
E		Ghana	1187	Iraq	216
Ecuador	2087	Greece	652	Ireland	1118
Egypt, Arab Rep.	51	Greenland		Isle of Man	
El Salvador	1724	Grenada	2350	Israel	435
Equatorial Guinea	2156	Guam		Italy	832
Eritrea	384	Guatemala	1996	**J**	
Estonia	626	Guinea	1651	Jamaica	2051
Ethiopia	848	Guinea-Bissau	1577	Japan	1668
F		Guyana	2387	Jordan	111
Faeroe Islands		**H**		**K**	
Fiji	2592	Haiti	1440	Kazakhstan	250
Finland	536	Honduras	1976	Kenya	630
France	867	Hong Kong SAR, China		Kiribati	
French Polynesia		Hungary	589	Korea, Dem. Rep	1054
				Korea, Rep	1274
				Kosovo	
				Kuwait	121

Country wise average precipitation in depth (mm per year) L-N

Country name	2011	Country name	2011	Country name	2011
L		**M**		**N**	
Lao PDR	1834	Macao SAR, China		Namibia	285
Latvia	641	Macedonia, FYR	619	Nepal	1500
Lebanon	661	Madagascar	1513	Netherlands	778
Lesotho	788	Malawi	1181	New Caledonia	
Liberia	2391	Malaysia	2875	New Zealand	1732
Libya	56	Maldives	1972	Nicaragua	2391
Liechtenstein		Mali	282	Nigeria	1150
Lithuania	656	Malta	560	Northern Mariana Islands	

(continued)

(continued)

Country name	2011	Country name	2011	Country name	2011
Luxembourg	934	Marshall Islands		Norway	1414
		Mauritania	92	**O**	
		Mauritius	2041	Oman	125
		Mexico	752	**P**	
		Micronesia, Fed. Sts.		Pakistan	494
		Moldova	450	Palau	
		Monaco		Panama	2692
		Mongolia	241	Papua New Guinea	3142
		Montenegro		Paraguay	1130
		Morocco	346	Peru	1738
		Mozambique	1032	Philippines	2348
		Myanmar	2091	Poland	600
				Portugal	854
				Puerto Rico	2054

Country wise average precipitation in depth (mm per year) Q-R

Country name	2011
Q	
Qatar	74
R	
Romania	637
Russian Federation	460
Rwanda	1212

Country wise average precipitation in depth (mm per year) S-Z

Country name	2011	Country name	2011	Country name	2011
S		**T**		**W**	
Samoa		Tajikistan	691	West Bank and Gaza	402
San Marino		Tanzania	1071	**X**	
Sao Tome and Principe	3200	Thailand	1622		
Saudi Arabia	59	Timor-Leste	1500	**Y**	
Senegal	686	Tajikistan	691	Yemen, Rep.	167
Serbia		Tanzania	1071	**Z**	
Seychelles	2330	Thailand	1622	Zambia	1020

(continued)

(continued)

Country name	2011	Country name	2011	Country name	2011
Sierra Leone	2526	Timor-Leste	1500	Zimbabwe	657
Singapore	2497	Togo	1168		
Sint Maarten (Dutch part)		Tonga			
Slovak Republic	824	Trinidad and Tobago	2200		
Slovenia	1162	Tunisia	207		
Solomon Islands	3028	Turkey	593		
Somalia	282	Turkmenistan	161		
South Sudan		Turks and Caicos Islands			
Spain	636	Tuvalu			
Sri Lanka	1712	U			
St. Kitts and Nevis	1427	Uganda	1180		
St. Lucia	2301	Ukraine	1875		
St. Martin (French part)		United Arab Emirates	78		
St. Vincent and the Grenadines	1583	United Kingdom	1220		
Sudan	416	United States	715		
Suriname	2331	V			
Swaziland	788	Vanuatu			
Sweden	624	Venezuela, RB	1875		
Switzerland	1537	Vietnam	1821		
Syrian Arab Republic	252	Virgin Islands (U.S.)			

Source http://data.worldbank.org/indicator/AG.LND.PRCP.MM

Table A.2 Rainfall intensity at different locations in Bangladesh

Months	Dhaka[a]		Chittagong[c]		Cox's Bazar		Barisal				Bangladesh 1971–2000[b]
	Av. Annual Rainfall (mm)	Rainy Days (Per Month)	Av. Annual Rainfall (mm)	Rainy Days	Av. Annual Rainfall (mm)	Rainy Days	Av. Annual Rainfall (mm)	Rainy Days	Av. Annual Rainfall (mm)	Rainy Days	Av. Annual Rainfall (mm)
Jan	4.5	1	6	2	19	–	18	–	66	–	7.53
Feb	22.1	2	28	3	15	–	18	–	40	–	22.88
Mar	58.1	4	63	5	44	–	98	–	43	–	50.54
Apr	153.5	8	151	7	168	–	90	–	51	–	121.56
May	282.8	12	265	13	254	–	143	–	67	–	277.32
Jun	385.7	17	533	22	309	–	189	–	153	–	457.68
Jul	364.9	20	598	26	436	–	159	–	132	–	530.33
Aug	302.0	17	519	25	249	–	192	–	92	–	430.99
Sep	289.3	14	321	21	212	–	158	–	261	–	320.94
Oct	159.8	8	180	8	174	–	105	–	78	–	169.04
Nov	36.5	2	55	2	122	–	59	–	33	–	47.51
Dec	10.3	1	16	1	13	–	43	–	36	–	10.72
Total			2735		735		487		358		

Sources

[a]Sonia Binte Murshed, A.K.M. Saiful Islam and M. Shah Alam Khan; 'Impact of climate change on rainfall intensity in Bangladesh'; 3rd International Conference on Water and Flood Management (ICWFM2011); http://teacher.buet.ac.bd/akmsaifulislam/publication/ICWFM2011_full_paper_119.pdf

[b]Mohammad Adnan Rajib, Md. Mujibur Rahman, A.K.M. Saiful Islam and Edward A. McBean; 'Analyzing the Future Monthly Precipitation Pattern in Bangladesh from Multi-Model Projections using both GCM and RCM'; World Environmental and Water Resources Congress 2011; Bearing Knowledge for Sustainability © ASCE 2011

[c]Rainfall/Precipitation in Chittagong, Bangladesh; http://www.chittagong.climatemps.com/precipitation.php Visited on 08 March 2014

Table A.3 Average annual rainfall of the states of India

Sl. no.	State	Meteorological Divisions	Average annual rainfall (mm)
1.	Andaman and Nicobar Islands	Andaman and Nicobar Islands	2967
2.	Arunachal Pradesh	Arunachal Pradesh	2782
3.	Assam	Assam and Meghalaya	2818
4.	Meghalaya	Assam and Meghalaya	2818
5.	Nagaland	Nagaland, Manipur, Mizoram and Tripura	1881
6.	Manipur	Nagaland, Manipur, Mizoram and Tripura	1881
7.	Mizoram	Nagaland, Manipur, Mizoram and Tripura	1881
8.	Tripura	Nagaland, Manipur, Mizoram and Tripura	1881
9.	West Bengal	n Sub-Himalayan West Bengal and Sikkim n Gangetic West Bengal	2739 1439
10.	Sikkim	Sub-Himalayan West Bengal and Sikkim	2739
11.	Orissa	Orissa	1489
12.	Bihar	n Bihar Plateau n Bihar Plains	1326 1186
13.	Uttar Pradesh	n Uttar Pradesh n Plain of West Uttar Pradesh n Hills of West Uttar Pradesh	1025 896 1667
14.	Haryana	Haryana, Chandigarh and Delhi	617
15.	Delhi	Haryana, Chandigarh and Delhi	617
16.	Chandigarh	Haryana, Chandigarh and Delhi	617
17.	Punjab	Punjab	649
18.	Himachal Pradesh	Himachal Pradesh	1251
19.	Jammu and Kashmir	Jammu and Kashmir	1011
20.	Rajasthan	n West Rajasthan n East Rajasthan	313 675
21.	Madhya Pradesh	n Madhya Pradesh n East Madhya Pradesh	1017 1338
22.	Gujarat	n Gujarat region n Saurashtra and Kachchh	1107 578
23.	Goa	Konkan and Goa	3005
24.	Maharashtra	n Konkan and Goa n Madhya Maharashtra n Marathwada n Vidarbha	3005 901 882 1034

(continued)

Table A.3 (continued)

Sl. no.	State	Meteorological Divisions	Average annual rainfall (mm)
25.	Andhra Pradesh	n Coastal Andhra Pradesh n Telengana n Rayalaseema	1094 961 680
26.	Tamil Nadu	Tamil Nadu and Pondicherry	998
27.	Pondicherry	Tamil Nadu and Pondicherry	998
28.	Karnataka	n Coastal Karnataka n North Interior Karnataka n South Interior Karnataka	3456 731 1126
29.	Kerala	Kerala	3055
30.	Lakshadweep	Lakshadweep	1515

Source http://www.rainwaterharvesting.org/urban/rainfall.htm visited on 20 May 2014

Table A.4 Rainfall and rainfall days of selected areas in Africa

1. Botswana (Maun)

	Jan	Feb	Mar	Apr	May	Jun	Jul	Aug	Sep	Oct	Nov	Dec
Rainfall (mm)	102	88	46	19	10	1	1	4	5	13	46	67
Rainfall (inch)	4.02	3.46	1.81	0.75	0.39	0.04	0.04	0.16	0.20	0.51	1.81	2.64
Rainfall days	14	10	8	3	1	0	0	0	1	3	8	11

2. Nirobi, Kenya

	Jan	Feb	Mar	Apr	May	Jun	Jul	Aug	Sep	Oct	Nov	Dec
Rainfall (mm)	74	49	89	120	129	16	15	30	21	37	151	79
Rainfall (inch)	2.91	1.93	3.50	4.72	5.08	0.63	0.59	1.18	0.83	1.46	5.94	3.11
Rainfall days	7	5	11	14	15	6	6	6	5	8	19	12

3. Windhoek, Namibia

	Jan	Feb	Mar	Apr	May	Jun	Jul	Aug	Sep	Oct	Nov	Dec
Rainfall (mm)	85	74	63	38	6	2	1	6	4	15	26	34
Rainfall (inch)	3.35	2.91	2.48	1.50	0.24	0.08	0.04	0.24	0.16	0.59	1.02	1.34
Rainfall days	15	14	13	7	1	0	0	1	2	4	7	9

4. Kigali, Rwanda

	Jan	Feb	Mar	Apr	May	Jun	Jul	Aug	Sep	Oct	Nov	Dec
Rainfall (mm)	54	111	78	180	87	9	3	21	72	111	90	75
Rainfall (inch)	2.13	4.37	3.07	7.09	3.43	0.35	0.12	0.83	2.83	4.37	3.54	2.95
Rainfall days	13	14	17	21	14	4	2	5	13	19	19	16

(continued)

Table A.4 (continued)

5. Johannesburg, South Africa

Rainfall (mm)	128	144	102	40	25	12	4	14	20	79	99	151
Rainfall (inch)	5.04	5.67	4.02	1.57	0.98	0.47	0.16	0.55	0.79	3.11	3.90	5.94
Rainfall days	19	15	15	10	5	3	1	3	5	15	18	21

6. Zanzibar, Tanzania

Rainfall (mm)	75	33	126	228	261	60	30	30	24	63	201	150
Rainfall (inch)	2.95	1.30	4.96	8.98	10.28	2.36	1.18	1.18	0.94	2.48	7.91	5.91
Rainfall days	12	7	15	19	16	8	7	9	8	11	16	17

7. Malawi (Lilongwe)

Rainfall (mm)	210	220	132	47	9	0	0	0	0	3	57	127
Rainfall (inch)	8.27	8.66	5.20	1.85	0.35	0	0	0	0	0.12	2.24	5.00
Rainfall days	20	17	15	6	3	2	1	0	1	1	4	8

http://www.eyesonafrica.net/africa-weather.htm

Table A.5 Average monthly rainfall (mm/month) in Sri Lanka

	Jan	Feb	Mar	Apr	May	Jun	Jul	Aug	Sep	Oct	Nov	Dec
Trincomalee	132	100	54	50	52	26	70	89	104	217	334	341
Nuwara Eliya	107	75	71	151	178	176	174	159	176	228	215	194
Colombo	62	69	130	253	382	186	125	114	236	369	310	168

Source http://www.roughguides.com/destinations/asia/sri-lanka/travel-essentials/average-monthly-temperatures-rainfall/

Table A.6 Availability of rainwater from flat surface (rainfall in mm/hr and area in sqm)

Rainfall	100	200	300	400	500	600	800	1000	1200	1400	1600	1800	2000
Surface area	Volume of rainwater available from flat surface in cum												
10	0.8	1.6	2.4	3.2	4	4.8	6.4	8	9.6	11.2	12.8	14.4	16
20	1.6	3.2	4.8	6.4	8	9.6	12.8	16	19.2	22.4	25.6	28.8	32
30	2.4	4.8	7.2	9.6	12	14.4	19.2	24	28.8	33.6	38.4	43.2	48
40	3.2	6.4	9.6	12	16	19.2	25.6	32	38.4	44.8	51.2	57.6	64
50	4	8	12	16	20	24	32	40	48	56	64	72	80
60	4.8	9.6	14.4	19.2	24	28.8	38.4	48	57.6	67.2	76.8	86.4	96
70	5.6	11.2	16.8	22.4	28	33.6	44.8	56	67.2	78.4	89.6	100.8	112
80	6.4	12.8	19.2	25.6	32	38.4	51.2	64	76.8	89.6	102.4	115.2	128
90	7.2	14.4	21.6	28.8	36	43.2	57.6	72	86.4	100.8	115.2	129.6	144
100	8	16	24	32	40	48	64	80	96144	112	128	144	160
150	12	24	36	48	60	72	96	120	192	168	192	216	240
200	16	32	48	64	80	96	128	160	240	224	256	288	320
250	20	40	60	80	100	120	160	200	240	280	320	360	400
300	24	48	72	96	120	144	192	240	288	336	384	432	480
400	32	64	96	128	160	192	256	320	384	448	512	576	640
500	40	80	120	160	200	240	320	400	480	560	640	720	800
1000	80	160	240	320	400	480	640	800	960	1120	1280	1440	1600
2000	160	320	480	640	800	960	1280	1600	1920	2240	2560	2880	3200
3000	240	480	720	960	1200	1440	1920	2400	2880	3360	3840	4320	4800

Source Central Public Works Department (CPWD), India 2002 'Rainwater Harvesting Conservation Manual'; Consultancy services, Organization, CPWD, New Delhi. Page 17
http://cpwd.gov.in/Publication/rain_wh.pdf dt. 25.03.14

Table A.7 Computation of peak flow of rainwater from a flat surface

Rainfall intensity mm/hr for 20 min.	50	100	150	200
Surface area	Pewak flow in lit/sec (lps)			
10	0.14	0.28	0.42	
20	0.28	0.56	0.83	1.11
30	0.42	0.83	1.25	1.67
40	0.56	1.11	1.67	2.22
50	0.69	1.39	2.08	2.78
60	0.83	1.67	2.50	3.33
70	0.97	1.94	2.92	3.89
80	1.11	2.22	3.33	4.44
100	1.39	2.78	4.17	5.55
200	2.78	5.56	8.33	11.11
500	6.95	13.89	20.83	27.78
1000	13.92	27.78	41.67	55.55

Source Ministry of water resources (MoWR) 2007 'Manual on artificial recharge of ground water'; Central ground water board, New Delhi, India. Page 128
http://cgwb.gov.in/documents/Manual%20on%20Artificial%20Recharge%20of%20Ground%20Water.pdf

Table A.8 Typical runoff coefficient

Materials	Note	Co-efficient
Pavement types		
Asphalt		0.95
Concrete		0.95
Brick		0.85
Gravel		0.75
Porous concrete unit pavers	<5 % slope	0.50
Porous concrete	<5 % slope	0.30
Porous asphalt	<5 % slope	0.40
Roof		
Asphalt		0.95
Green roof	Depth <4 in.	0.50
	Depth 4–8 in.	0.30
	Depth 9–20 in.	0.20
	Depth >20 in.	0.10
Turf		
Flat	0–1 % slope	0.25
Average	1–3 % slope	0.30
Hilly	3–10 % slope	0.35
Steep	>10 % slope	0.40
Vegetation		
Flat	0–1 % slope	0.10
Average	1–3 % slope	0.20
Hilly	3–10 % slope	0.25
Steep	>10 % slope	0.30
Filtration basin		0.00

Source http://www.discoverdesign.org/discover/math/soil-permeability

Table A.9 National standard taper pipe threads size NPT chart

Outside diameter D	Nominal pipe size	Threads per inch (Pitch)	Thread pitch length, P	Pitch diameter at external thread start E0	Hand tight thread engagement length, L1	Diameter E1	Maximum thread engagement L2	Diameter E2
0.3125	1/16	27	0.03704	0.27118	0.160	0.28118	0.2611	0.28750
0.405	1/8	27	0.03704	0.36351	0.1615	0.37360	0.2639	0.38000
0.540	1/4	18	0.05556	0.47739	0.2278	0.49163	0.4018	0.50250
0.675	3/8	18	0.05556	0.61201	0.240	0.62701	0.4078	0.63750
0.840	1/2	14	0.07143	0.75843	0.320	0.77843	0.5337	0.79179
1.050	3/4	14	0.07143	0.96768	0.339	0.98887	0.5457	1.00179
1.315	1	11½	0.08696	1.21363	0.400	1.23863	0.6828	1.25630
1.660	1¼	11½	0.08696	1.55713	0.420	1.58338	0.7068	1.60130
1.900	1 1/2	11½	0.08696	1.79609	0.420	1.82234	0.7235	1.84130
2.375	2	11½	0.08696	2.26902	0.436	2.29627	0.7565	2.31630
2.875	2½	8	0.12500	2.71953	0.682	2.76216	1.1375	2.79062
3.500	3	8	0.12500	3.34062	0.766	3.38850	1.2000	3.41562
4.000	3½	8	0.12500	3.83750	0.821	3.88881	1.2500	3.91562
4.500	4	8	0.12500	4.33438	0.844	4.38712	1.3000	4.41562
5.563	5	8	0.12500	5.39073	0.937	5.44929	1.4063	5.47862
6.625	6	8	0.12500	6.44609	0.958	6.50597	1.5125	6.54062
8.625	8	8	0.12500	8.43359	1.063	8.50003	1.7125	8.54062
10.750	10	8	0.12500	10.54531	1.210	10.62094	1.9250	10.66562
12.750	12	8	0.12500	12.53281	1.360	12.61781	2.1250	12.66562
14.000	14	8	0.12500	13.77500	1.562	13.87262	2.2500	13.91562
16.000	16	8	0.12500	15.76250	1.812	15.87575	2.4500	15.91562
18.000	18	8	0.12500	17.75000	2.000	17.87500	2.6500	17.91562
20.000	20	8	0.12500	19.73750	2.125	19.87031	2.8500	19.91562
24.000	24	8	0.12500	23.71250	2.375	23.86094	3.2500	23.91562

Reference ANSI/ASME B1.20.1-1983 (R1992) Angle between sides of thread is 60 °. Thread taper, is 3/4 in. per foot. All units are in inches.

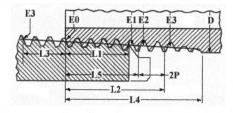

Source http://www.engineersedge.com/hardware/taper-pipe-threads.htm

Table A.10 uPVC pressure (water supply, irrigation and industrial use)

Nominal Size (mm)	Outside Diameter (mm)		Wall thickness (mm)									
			Class "B"		Class "C"		Class "D"		Class "E"		Class "O"	
	Min	Max	Min	Max	Min	Max	Min	Max	Min	Max	Min	Max
12	21.2	21.5										
19	26.6	26.9										
25	33.4	33.7										
32	42.1	42.4					2.2	2.7	2.7	3.2		
40	48.1	48.4					2.5	3.0	3.1	3.7	1.8	2.2
50	60.2	60.5			2.5	3.0	3.1	3.7	3.9	4.5	1.8	2.2
63	75.0	75.3			3.0	3.5	3.9	4.5	4.8	5.5	1.8	2.2
76	88.7	89.1	2.9	3.4	3.5	4.1	4.6	5.3	5.7	6.6	1.8	2.2
110	114.1	114.5	3.4	4.0	4.5	5.2	6.0	6.9	7.3	8.4	2.3	2.8
160	168.0	168.5	4.5	5.2	6.6	7.6	8.8	10.2	10.8	12.5	3.1	3.7
200	218.8	219.4	5.3	6.1	7.8	9.0	10.3	11.9	12.6	14.5	3.1	3.7

Pressure Ratings: Designated by the different classes at 20 °C

Class	'B'	'C'	'D'	'E'	'O'
Bar	6	9	12	15	Non Pressure

Note
2 % of rated pressure should be reduced for each 1 °C rise above 20 °C
Manufactured to : BS 3505/3506 Classes B, C, D and E, BS 3506, 1969 Class O
Standard Length : 5.8 and 6 m
Colour: Dark grey (except Class O which is grey)
Socket type : Solvent weld/Plain end
Source file:///C:/Users/User/Downloads/british%20stardard.pdf

Table A.11 Drinking water quality standards

Parameter	World Health Organization	European Union	EPA, United States
1,2 dichloroethane		3.0 µg/l	5 µg/l
Acrylamide		0.10 µg/l	
Antimony	ns	5.0 µg/l	6.0 µg/l
Arsenic	10 µg/l	10 µg/l	10 µg/l
Barium	700 µg/l	ns	2 mg/L
Benzene	10 µg/l	1.0 µg/l	5 µg/l
Benzo(a) pyrene		0.010 µg/	0.2 µg/l
Boron	2.4 mg/l	1.0 mg/L	
Bromate		10 µg/l	10 µg/l
Cadmium	3 µg/l	5 µg/l	5 µg/l
Chromium	50 µg/l	50 µg/l	0.1 mg/L
Copper		2.0 mg/l	TT
Cyanide		50 µg/l L	0.2 mg/L
Epichlorohydrin		0.10 µg/l	
Fluoride	1.5 mg/l	1.5 mg/l	4 mg/l
Lead		10 µg/l	15 µg/l
Mercury	6 µg/l	1 µg/l	2 µg/l
Nickel		20 µg/l	
Nitrate	50 mg/l	50 mg/l	10 mg/L (as N)
Nitrite		0.50 mg/l	1 mg/L (as N)
Pesticides (individual)		0.10 µg/l	
Pesticides (Total)		0.50 µg/l	
Polycyclic aromatic hydrocarbons l		0.10 µg/	
Selenium	40 µg/l	10 µg/l	50 µg/l
Tetrachloroethene and Trichloroethene	40 µg/l	10 µg/l	

Source Wikipedia, the free encyclopedia http://en.wikipedia.org/wiki/Drinking_water_quality_standards

Table A.12 Bangladesh drinking water quality standards

Sl. No.	Water quality parameter	Unit	Bangladesh Standard
01.	Aluminium	mg/l	0.2
02.	Ammonia	mg/l	0.5
03.	Arsenic	mg/l	0.05
04.	Barium	mg/l	0.01
05.	Benzene	mg/l	0.01
06.	BOD5	mg/l	0.2

(continued)

Table A.12 (continued)

Sl. No.	Water quality parameter	Unit	Bangladesh Standard
07.	Boron	mg/l	1.0
08.	Cadmium	mg/l	0.005
09.	Calcium	mg/l	75
10.	Chloride	mg/l	150–600
11.	Chlorinated alkenes		
	Carbon tetrachloride	mg/l	0.01
	1,1 Dichloroethylene	mg/l	0.001
	1,2 Dichloroethylene	mg/l	0.03
	Tetrachloroethylene	mg/l	0.03
	Trichloroethylene	mg/l	0.09
12.	Chlorinated phenol		
	Pentachlorophenol	mg/l	0.03
	2,4,6 Trichlorophenol	mg/l	0.03
13.	Chlorine (residual)	mg/l	0.2
14.	Chloroform	mg/l	0.09
15.	Chromium (hexavalent)	mg/l	0.05
16.	Chromium (total)	mg/l	0.05
17.	COD	mg/l	4
18.	Coliform (faecal)	mg/l	0
19.	Coliform (total)	mg/l	0
20.	Color	mg/l	15
21.	Copper	mg/l	1
22.	Cyanide	mg/l	0.1
23.	Detergent	mg/l	0.2
24.	Dissolve oxygen	mg/l	6
25.	Flouride	mg/l	1
26.	Hardness (as $CaCO_3$)	mg/l	200–500
27.	Iron	mg/l	0.3–1.0
28.	Kjehldal Nitrogen	mg/l	1
29.	Lead	mg/l	0.05
30.	Magnesium	mg/l	30–35
31.	Manganese	mg/l	0.1
32.	Mercury	mg/l	0.001
33.	Nickel	mg/l	0.1
34.	Nitrate	mg/l	10
35.	Nitrite	mg/l	<1.0
36.	Odor	mg/l	Odorless
37.	Oil & grease	mg/l	0.01
38.	pH	mg/l	6.5–8.5
39.	Phenolic compounds	mg/l	0.002
40.	Phosphate	mg/l	6
41.	Phosphorus	mg/l	0

(continued)

Table A.12 (continued)

Sl. No.	Water quality parameter	Unit	Bangladesh Standard
42.	Potassium	mg/l	12
43.	Radioactive substance	mg/l	
	Total alfa radiation	mg/l	0.01
	Total beta radiation	mg/l	0.1
44.	Selenium	mg/l	0.01
45.	Silver	mg/l	0.02
46.	Sodium	mg/l	200
47.	Suspended solid	mg/l	10
48.	Sulfide	mg/l	0
49.	Sulfate	mg/l	400
50.	Total Dissolve Solid (TDS)	mg/l	1000
51.	Temperature	mg/l	20–30
52.	Tin	mg/l	2
53.	Turbidity	mg/l	10
54.	Zinc	mg/l	5

Source http://www.amiowater.com/index.php?menu_id=53&exmenu=53&page=1

Table A.13 Pre-fabricated water tanks (capacity and dimension)

A: Square look modular tanks

Capacity (Approximate)		Length		Width		Height	
Modules	Liters	mm	foot	mm	foot	mm	foot
1	400	600	2'-0"	600	2'-0"	1140	3'-9"
2	550	1140	3'-9"	550	1'-10"	840	2'-9"
2	750	1140	3'-9"	550	1'-10"	1170	3'-10"
3	850	1740	5'-9"	550	1'-10"	840	2'-9"
3	1150	1740	5'-9"	550	1'-10"	1170	3'-10"
4	1500	2340	7'-8"	550	1'-10"	1140	3'-9"
3	1650	1740	5'-9"	800	2'-8"	1140	3'-9"
4	2250	2340	7'-8"	800	2'-8"	1140	3'-9"

B: Circular modular tanks

Capacity (Liter)	Diameter		Height		Manhole	
	(mm)	Foot	(mm)	Foot	(mm)	Foot
500	1000	3'-4"	735	2'-5.5"	400	1'-4'
1000	1115	3'-8.5"	1115	3'-8.5"	400	1'-4"
2000	1390	4'-7.5"	1500	5'-0"	480	1'-7.2"
3000	1640	5'-5.5"	1675	5'-7"	480	1-7.2"
5000	2000	6'-8"	1880	6'-3.2"	480	1-7.2"
10,000	2300	7'-8"	3100	10'-4"	480	1-7.2"

Source http://www.denyertanks.com.au/wp/wp-content/uploads/2013/07/table2.gif
Source http://www.bharattradingco.com/water-tank.htm

Table A.14 Estimating water demand against Water Supply Fixture Unit (WSFU)

Supply systems predominantly for flush tanks			Supply systems predominantly for flush valves		
Load	Demand		Load	Demand	
(WSFU)	(Gallons per minute)	(Cubic feet per minute)	(WSFU)	(Gallons per minute)	(Cubic feet per minute)
1	3.0	0.04104	–	–	–
2	5.0	0.0684	–	–	–
3	6.5	0.86892	–	–	–
4	8.0	1.06944	–	–	–
5	9.4	1.256592	5	15.0	2.0052
6	10.7	1.430376	6	17.4	2.326032
7	11.8	1.577424	7	19.8	2.646364
8	12.8	1.711104	8	22.2	2.967696
9	13.7	1.831416	9	24.6	3.288528
10	14.6	1.951728	10	27.0	3.60936
11	15.4	2.058672	11	27.8	3.716304
12	16.0	2.13888	12	28.6	3.823248
13	16.5	2.20572	13	29.4	3.930192
14	17.0	2.27256	14	30.2	4.037136
15	17.5	2.3394	15	31.0	4.14408
16	18.0	2.90624	16	31.8	4.241024
17	18.4	2.459712	17	32.6	4.357968
18	18.8	2.513184	18	33.4	4.464912
19	19.2	2.566656	19	34.2	4.571856
20	19.6	2.620128	20	35.0	4.6788
25	21.5	2.87412	25	38.0	5.07984
30	23.3	3.114744	30	42.0	5.61356
35	24.9	3.328632	35	44.0	5.88192
40	26.3	3.515784	40	46.0	6.14928
45	27.7	3.702936	45	48.0	6.41664
50	29.1	3.890088	50	50.0	6.684
60	32.0	4.27776	60	54.0	7.21872
70	35.0	4.6788	70	58.0	7.75344
80	38.0	5.07984	80	61.2	8.181216
90	41.0	5.48088	90	64.3	8.595624
100	43.5	5.81508	100	67.5	9.0234
120	48.0	6.41664	120	73.0	9.75864
140	52.5	7.0182	140	77.0	10.29336
160	57.0	7.61976	160	81.0	10.82808
180	61.0	8.15448	180	85.5	11.42964

http://publicecodes.cyberregs.com/st/sc/b9v07/st_sc_st_b9v07_appe_par013.htm

Table A.15 Pressure loss in fittings and valves expressed as equivalent length of pipe (feet)

Nominal or standard size (inches)	FITTINGS				Coupling	VALVES			
	Standard EII		90° Tee			Ball	Gate	Butterfly	Check
	90°	45°	Side Branch	Straight Run					
3/8	0.5	–	1.5	–	–	–	–	–	1.5
1/2	1	0.5	2	–	–	–	–	–	2
5/8	1.5	0.5	2	–	–	–	–	–	2.5
3/4	2	0.5	3	–	–	–	–	–	3
1	2.5	1	4.5	–	–	0.5	–	–	4.5
1¼	3	1	5.5	0.5	0.5	0.5	–	–	5.5
1½	4	1.5	7	0.5	0.5	0.5	–	–	6.5
2	5.5	2	9	0.5	0.5	0.5	0.5	7.5	9
2½	7	2.5	12	0.5	0.5	–	1	10	11.5
3	9	3.5	15	1	1	–	1.5	15.5	14.5
3½	9	3.5	14	1	1	–	2	–	12.5
4	12.5	5	21	1	1	–	2	16	18.5
5	16	6	27	1.5	1.5	–	3	11.5	23.5
6	19	7	34	2	2	–	3.5	13.5	26.5
8	29	11	50	3	3	–	5	12.5	39

Ref: ICC International Code Council, South Carolina Codes 2007
http://publicecodes.cyberregs.com/st/sc/b9v07/st_sc_st_b9v07_appe_par013.htm

See Chart A.1.

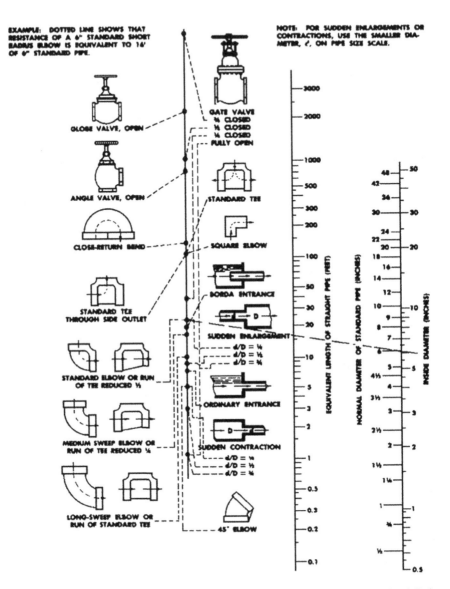

Chart A.1 Friction effects of fittings and of changes in pipe sizes. Ref: ICC International Code Council, South Carolina Codes 2007. http://publicecodes.cyberregs.com/st/sc/b9v07/st_sc_st_b9v07_appe_par013.htm

Table A.16 British standard 3505: 1968 and 1986 of uPVC pipe (for cold rainwater supply)

Pipe dimensions						
Nominal Size (inch)	Mean outside diameter	Wall thickness				
		Class O (non-pressure)	Class B over 6 Bar	Class C over 9 bar	Class D 12 Bar	Class E 15 Bar
	min (mm)	averaged value	averaged value	averaged value	averaged value	averaged value
		max (mm)	max (mm)	max (mm)	max (mm)	max (mm)
3/8	17.0					1.9
1/2	21.2					2.1
3/4	26.6					2.5
1	33.4					2.7
1¼	42.1				2.7	3.2
1½	48.1	2.2			3.0	3.7
2	60.2	2.2	3.0	3.0	3.7	4.5
2½	75.0	2.2		3.5	4.5	5.5
3	88.7	2.2	3.4	4.1	5.3	6.5
4	114.1	2.8	3.8	5.2	6.8	8.3
5	140.0	3.1	4.4	6.3	8.3	10.1
6	168.0	3.7	5.2	7.5	9.9	12.1
7	193.5	3.7	6.0	8.7	11.4	13.9
8	218.8	3.7	6.1	8.8	11.6	14.1

Table A.17 Infiltration rates for different type soils

Soil type	Porosity (percent)		Infiltration rate		
	Total	Non-capillary (specific) yield	Inches per hour	Feet per day	Centimeters per hour
Gravelly silt loam	54.9	28.1	4.96	9.92	12.60
Clay loam	61.1	36.3	3.98	7.96	10.11
Silt loam	57.0	32.0	2.09	4.18	5.31
Sandy loam	49.6	26.3	1.93	3.86	4.90
Clay (eroded)	54.3	28.7	1.78	3.56	4.52
Sandy clay loam	48.8	27.7	1.42	2.84	3.61
Silty clay loam	50.8	24.3	0.72	1.44	1.83
Stony silt loam	59.7	32.6	0.55	1.10	1.40
Fine sandy loam	41.5	24.2	0.55	1.10	1.40
Very fine sandy loam	49.6	23.4	0.51	1.02	1.29
Loam	45.7	17.2	0.50	1.00	1.27

(continued)

Table A.17 (continued)

Soil type	Porosity (percent)		Infiltration rate		
	Total	Non-capillary (specific) yield	Inches per hour	Feet per day	Centimeters per hour
Sandy clay	42.9	16.9	0.05	0.10	0.13
Heavy clay	57.8	27.0	0.02	0.04	0.05
Light clay	47.0	19.8	0.00	0.00	0.00
Claye silt loam	49.4	17.6	0.00	0.00	0.00

Measured by infiltrometer rings in third hour of a wet run [After Free, Browning, and Musgrave, (1940)]

Table A.18 Maximum allowable flow velocities for open soil (non-vegetated) low flow channels

Soil description		Allowable flow velocity (m/s)
Extremely erodible soils		0.3
Highly erodible soils (black earth, fine surface texture soils)		0.5
Moderately erodible soils		0.6
Low erodible soils (krasnozems, red earth)		0.7
Sandy soils (Manning's $n = 0.04$)		0.45
Fine colloidal sand ($n = 0.02$)		0.45
Sandy loam, non-colloidal ($n = 0.02$)		0.5
Alluvial silts or silt loam, non-colloidal ($n = 0.02$)		0.6
Fine gravel or firm loam ($n = 0.02$)		0.7
Graded loam to cobble, non-colloidal ($n = 0.03$)		1.1
Alluvial silts, colloidal ($n = 0.025$)		1.1
Stiff clay, very colloidal ($n = 0.025$)		1.1
Coarse gravel, non-colloidal ($n = 0.025$)		1.2
Graded silts to cobbles when colloidal ($n = 0.03$)		1.2
Loose rock, nominal size around 200 mm		1.5
Cohesive soils	Lean clayey soils	Heavy clayey soils
Loose	0.34	0.46
Fairly compact	0.6	0.7
Compact	0.9	1.0
Very compact	1.2	1.5

Source http://www.brisbane.qld.gov.au/sites/default/files/ncd_sect3_design_part7.pdf

Table A.19 Conversion factors, mixed units

Length (L)

mile	yard	ft	in.	m	cm	
1	1760	5280	6.336×10^4	1.609×10^3	1.609×10^5	
5.68×10^{-4}	1	3	36	0.9144	91.44	
1.894×10^{-4}	0.333	1	12	0.3048	30.48	
1.578×10^{-5}	0.028	0.083	1	0.0254	2. 54	
6.214×10^{-4}	1.094	3.281	39.37	1	100	

Area (A)

mile^2	acre	yard^2	ft^2	in^2	m^2	
1	640	3.098×10^6	2.788×10^7	4.014×10^9	2.59×10^6	
1.563×10^{-3}	1	4840	43,560	6.27×10^6	4047	
3.228×10^{-7}	2.066×10^{-4}	1	9	1296	0.836	
3.587×10^{-8}	2.3×10^{-5}	0.111	1	144	0.093	
2.491×10^{-10}	1.59×10^{-7}	7.716×10^{-4}	6.944×10^{-3}	1	6.452×10^{-4}	
3.861×10^{-7}	2.5×10^{-4}	1.196	10.764	1550	1	

Volume (V)

acre-ft	U.S. Gal	ft^3	in.^{3}	L	m^3	cm^3
1	325,851	43,560	75.3×10^{-6}	1.23×10^6	1230	1.23×10^9
3.07×10^{-6}	1	0.134	231.6	3.875	3.875×10^{-3}	3875
2.3×10^{-5}	7.481	1	1728	28.317	0.028	28,317
1.33×10^{-8}	4.329×10^{-3}	5.787×10^{-4}	1	0.016	1.639×10^{-5}	16.39
8.1×10^{-7}	0.264	0.035	61.02	1	1×10^{-3}	1000
8.13×10^{-4}	264.2	35.31	6.10×10^4	1000	1	10^6

Time (T)

year	month	day	hour	minute	second
1	12	365	8760	525,600	3.1536×10^7

Velocity (L/T)

ft/s	ft/min	m/s	m/min	cm/s	
1	60	0.3048	18.29	30.48	
0.017	1	5.08×10^{-3}	0.3048	0.5080	
3.281	196.8	1	60	100	
0.055	3.28	0.017	1	1.70	
0.032	1.969	0.01	0.588	1	

Discharge (L^3/T)

mgd	gpm	ft^3/s	ft^3/min	L/s	m^3/d
1	694.4	1.547	92.82	43.75	3.78×10^3
1.44×10^{-3}	1	2.228×10^{-3}	0.134	0.063	5.45
0.646	448.9	1	60	28.32	2447
0.011	7.481	0.017	1	0.472	40.78

(continued)

Table A.19 (continued)

Discharge (L^3/T)

mgd	gpm	ft³/s	ft³/min	L/s	m³/d
0.023	15.85	0.035	2.119	1	86.41
2.64×10^{-4}	0.183	4.09×10^{-4}	0.025	0.012	1

Mass (M)

ton	lb_m	grain	ounce (oz)	kg	g
1	2000	1.4×10^7	32,000	907.2	907,185
0.0005	1	7000	16	0.454	454
7.14×10^{-8}	1.429×10^{-4}	1	2.29×10^{-3}	6.48×10^{-5}	0.065
3.125×10^{-5}	0.0625	437.6	1	0.028	28.35
1.10×10^{-3}	2.205	1.54×10^4	35.27	1	1000
1.10×10^{-6}	2.20×10^{-3}	15.43	0.035	10^{-3}	1

Temperature (T)

°F	°C	°K	°R
°F	5/9(°F − 32)	5/9 °F + 255.38	°F + 459.69
9/5 °C + 32	°C	°C + 273.16	9/5 °C + 491.69
9/5 °K − 459.69	°K − 273.16	°K	9/5 °K
°R − 459.69	5/9°R − 273.16	5/9°R	°R

Density (M/L^3)

lb/ft³	lb/gal(U.S)	kg/m³	kg/L	g/cm³
1	0.1337	16.019	0.01602	0.01602
7.48	1	119.8	0.1198	0.1198
0.0624	8.345×10^{-3}	1	0.001	0.001
62.43	8.345	1000	1	1

Pressure (F/L^2)

lb/in²	ft water	in Hg	atm	mm Hg	kg/cm²	N/m²
1	2.307	2.036	0.068	51.71	0.0703	6895
0.4335	1	0.8825	0.0295	22.41	0.0305	2989
0.4912	1.133	1	0.033	25.40	0.035	3386
14.70	33.93	29.92	1	760	1.033	1.013×10^5
0.019	0.045	0.039	1.30×10^{-3}	1	1.36×10^{-3}	133.3
14.23	32.78	28.96	0.968	744.7	1	98,070
1.45×10^{-4}	3.35×10^{-4}	2.96×10^{-4}	9.87×10^{-6}	7.50×10^{-3}	1.02×10^{-5}	1

Viscosity

Dynamic absolute viscosity (μ)

cp	lbf.ft2	lb_m/ft.s	g/cm.s	N.s/m²	kg/m.s	dp
1	2.09×10^{-5}	6.72×10^{-4}	0.01	1×10^{-3}	1×10^{-3}	1×10^{-3}
4.78×10^4	1	32.15	478.5	47.85	47.85	47.85
1488	0.031	1	14.88	1.488	1.488	1.488

(continued)

Table A.19 (continued)

Viscosity

Dynamic absolute viscosity (μ)

cp	Ibf.ft2	Ib_m/ft.s	g/cm.s	N.s/m^2	kg/m.s	dp
100	2.09×10^{-3}	0.672	1	0.10	0.10	0.10
1000	0.021	0.672	10	1	1	1

Kinemate Viscosity (V)

centistoke	ft^2/s	cm^2/s	m^2/s	Myriastoke
1	1.076×10^{-5}	0.01	1.0×10^{-6}	1.0×10^{-6}
9.29×10^4	1	929.4	0.093	0.093
100	1.076×10^{-3}	1	1.0×10^{-4}	1.0×10^{-4}
10^6	10.76	10^4	1	1

Force (F)

Ib$_f$	N	dyne
1	4.448	4.448×10^5
0.225	1	10^5
2.25×10^{-6}	10^{-5}	1

Energy (E)

kW.h	hp.h	Btu	J	kj	calories
1	1.341	3412	3.6×10^6	3600	8.6×10^5
0.7457	1	2545	2.684×10^6	2685	6.4×10^5
2.930×10^{-6}	3.929×10^{-4}	1	1055	1.055	252
2.778×10^{-7}	3.72×10^{-7}	9.48×10^{-4}	1	0.001	0.239
2.778×10^{-4}	3.72×10^{-4}	0.948	1000	1	239
1.16×10^{-6}	1.56×10^{-6}	3.97×10^{-3}	4.186	4.18×10^{-3}	1

Power (P)

kW	Btu/min	hp	ft.b/s	kg.m/s	cal/min
1	56.89	1.341	737.6	102	14,330
0.018	1	0.024	12.97	1.793	252
0.746	42.44	1	550	76.09	10,690
1.35×10^{-3}	0.077	1.82×10^{-3}	1	0.138	19.43
9.76×10^{-3}	0.558	0.013	7.233	1	137.6
6.98×10^{-5}	3.97×10^{-3}	9.355×10^{-5}	0.0514	7.12×10^{-3}	1

Source Quasim (2002)

Index

A
ABS pipe, 194
Acid, 40, 113, 193, 209, 229
Aggregate, 83, 169, 225, 227–229
Air release valve, 215
Angle globe valve, 215
Angle stop cock, 221
Asphalt shingle, 82, 83

B
Backfill, 184, 232, 233
Bacteria, 8, 39, 67, 82, 114, 124, 126–128
Ball valve, 215
Bathing water, 7, 61, 67, 69, 107, 118, 136
Bathroom, 63
Bedding, 184, 232
Boiling, 126–128
Boulder, 162, 164, 169, 192
Brick, 83, 161, 162, 167, 180, 192, 209,
    225–227, 229, 233, 234
Building, 1, 2, 4–6, 8–10, 12–18, 23, 29, 38,
    45–53, 55, 57, 59–71, 73–76, 80, 87, 88,
    90, 94, 100–104, 106, 108–110, 112–115,
    126, 128, 132, 135, 140, 150, 153, 159,
    161, 164, 173, 176–178, 180, 183, 194,
    208, 209, 211, 221, 227
Bushing, 198, 199
Bus station, 62, 63
Butterfly valve, 215
Butt joint, 195, 205

C
Campylobacter, 39
Cap, 201, 206
Catchment, 1, 5, 9, 38–40, 48, 51–56, 73–75,
    80, 83–91, 93, 94, 102, 103, 110, 111, 118,
    119, 153, 161, 162, 164, 168, 176, 183, 192
Cement, 40, 73, 110, 161, 162, 192, 194, 195,
    198, 203, 205, 208, 209, 225–230, 233

Cement–sand mortar, 228
Check valve, 136, 137, 213, 215, 217
Chlorination, 126, 129, 131
Clay tile, 110
Clothe washing, 61
Cloud, 4, 23–25, 31, 40
Cock, 206, 215, 221, 222
Collar, 198
Compression faucet, 221
Concrete, 29, 40, 51, 73, 82, 95, 167, 181, 184,
    185, 188, 195, 205, 208, 213, 226–230,
    233, 234
Concrete pipe, 184, 185, 192, 195, 196, 205,
    232
Concrete tile, 82, 162, 163
Constructed tanks, 114, 115, 208
Coupling, 146, 196, 197, 213
CPVC pipe, 194
Crosses, 199, 200
Cross section, 88, 91, 96, 179, 181, 182, 184,
    213
Cryptosporidium, 39

D
Delivery head, 141, 143
Depth, 4, 25–27, 37, 84, 88, 89, 105, 111, 122,
    123, 155, 158, 162, 164, 167, 170, 182,
    187, 188, 232, 233
Detention period, 122, 123
Direct-pumping, 135, 137, 138, 140
Discharge capacity, 179, 180
Disinfection, 5, 12, 45, 50, 118, 126–132
Drainage pipe, 52, 114, 115, 161, 183, 184,
    232
Drain cock, 221
Drinking water, 5, 7–10, 130
Dry weather flow, 181

© Springer International Publishing Switzerland 2017
S.A. Haq, PEng, *Harvesting Rainwater from Buildings*,
DOI 10.1007/978-3-319-46362-9

E

Effective porosity, 156–158
Efficiency, 12, 51, 123, 146, 181, 218, 231
Elbow, 199
Escherichia coli, 39, 56, 126, 127

F

Factory, 63, 65, 122, 164, 213, 223, 229
Faucet, 64, 67, 71, 102, 137, 144, 147, 148, 206, 215, 221–224
Feeler gauge, 213
Ferro-cement tank, 209
Filtering gravel, 231
Filtering sand, 124, 231
Filter media, 122, 124, 125, 165, 168, 192, 231
Filtration, 12, 45, 50, 118, 122–126, 132, 231
Fire, 5, 60, 64, 65, 118, 214
First flush–diversion, 92
Fittings, 52, 113, 141, 144, 194, 196, 198, 199, 201, 203
Fixture unit, 144, 145, 148–151
Flange, 201
Flap valve, 217
Floating ball, 93
Float valves, 218
Flow, 2, 65, 67, 70, 82–84, 88, 91, 96, 112, 114, 119–124, 141, 143, 144, 148, 149, 157, 170, 176, 180–182, 215, 223, 225
Flushing toilet, 11, 61
Fountain, 5, 61, 62, 65, 66, 69, 118, 145
Freeboard, 89, 183
French drains, 178
Fuller faucet, 221, 223

G

Galvanized Iron pipe, 192
Gate valve, 92, 93, 136, 137, 213, 215, 216
Gauge, 23, 27–30, 213
Giardia, 39, 127
G.I. tank, 210
Globe valve, 215
Grading, 185, 186, 232
Gravel, 3, 110, 123–125, 158, 161, 162, 164, 167–170, 180, 192, 227, 231, 232
Grouting, 213
Gutter, 7, 18, 48, 53, 56, 79, 80, 83–85, 88, 89, 96, 162

H

Head, 124, 137–139, 141–144, 147, 156, 158, 205
Hospital, 62, 63
Hostel, 63
House washing, 61

I

Imperviousness, 175
Infiltration, 4, 27, 156, 161–163, 165, 167, 169
Infiltration rate, 155–157, 162, 165
Injection well, 161, 169–171
Inspection pit, 161, 169, 186–226, 233

J

Jar, 7, 29, 30

K

Kinetic head, 143

L

Leveling, 213
Lift-check valve, 217
Lime terrace, 82

M

Manhole cover, 187, 192, 225, 233–235
Manholes, 114, 174, 186–188, 233–235
Masonry tanks, 113, 208, 209
Mixture faucet, 221, 223

N

Nipple, 196, 197

O

Office, 62–64, 68
Offset, 90, 199, 200
On–off valve, 215, 216
Overflow rate, 120, 121, 123
Ozonation, 126, 131
Ozone, 126, 131

P

Paint, 82, 194
Park, 62, 67, 175, 235
Peak runoff, 176, 178, 179
Permeability, 156, 158, 164
Pipe, 13, 18, 39, 52, 55, 56, 68, 74, 76, 83, 84, 86, 88–93, 96, 102, 111, 113–115, 125, 136, 138, 141, 144, 147–151, 160, 168, 170, 174, 183–185, 187, 188, 192–196, 198, 199, 201–207, 213, 215, 217–219, 221, 223, 224, 232–234
Pipe jointing, 202–204
Pit, 161, 162, 164, 168, 225, 229, 230, 233, 234
Plain faucet, 221–223
Plastic pipe, 192–194, 203, 232
Plastic tank, 211
Plug, 201
Potential head, 143

Power, 129, 138, 141, 146, 147, 212, 214
Pressure reducing valve, 218
Pseudomonas, 39
Pump, 52, 55, 105, 126, 140, 141, 143, 146,
    147, 170, 174, 192, 212–214
Pump room, 213

R
Railway station, 62
Rain, 2, 6, 12, 23–26, 28, 29, 31, 32, 56, 74,
    86, 90, 92, 105
Rain gauge, 23, 27–29, 31
Rainwater, 1, 2, 4–17, 19, 23, 25, 27, 28, 30,
    37–40, 45–47, 49–51, 53, 54, 57, 59, 60,
    64, 66–68, 73–76, 80, 82, 83, 85, 87–90,
    92–96, 99–103, 105–111, 114, 115,
    117–120, 122, 123, 125–127, 129, 130,
    135, 136, 138, 141, 149, 150, 153–155,
    157, 159, 161, 162, 164, 166, 168, 170,
    173, 174, 177, 183, 191–193, 202, 207,
    208, 215, 221, 224, 231
Rainwater down pipe (RDP), 84, 87–90
Recharge, 8, 9, 47, 52, 154–156, 159, 162, 164,
    165, 168–170, 225, 226
Recharge pit, 161, 162, 164
Recharge trench, 161, 164
Recharge troughs, 161, 164
Recharge well, 161, 165–170
Recreational Water, 65
Reducers, 198
Reinforced cement concrete (RCC), 195, 198,
    209, 230
Reinforcing bar, 192, 209, 225, 227, 228, 230
Retention time, 156, 157
Roof, 77–81
    butterfly roof, 79, 80
    gambrel roof, 77, 79
    gazebo roof, 77, 78
    hipped roof, 77
    mansard roof, 77, 78
    M-shaped roof, 79, 80
    pitch or gable roof, 77, 78
    pyramid hip roof, 77
    saw-tooth roof, 79
    shed roof, 77, 78
Roughness coefficient, 179, 180
Runoff coefficient, 110, 179

S
Salmonella, 39, 127
Salt, 25, 32, 40, 154, 209
Sand, 3, 119, 123–125, 162–164, 167, 170,
    192, 225, 227–229, 231, 232

School, 62, 63
Screening, 118, 119
Sedimentation, 45, 49, 52, 55, 118, 120, 122,
    123, 168
Self closing faucet, 221, 222
Sewer, 61, 62, 64, 69, 104, 159, 176, 177, 181,
    183
Shigella, 39, 127
Shopping center, 62, 63
Slate, 83, 110
Spigot joint, 205
Steel tank, 210
Storage tank, 7, 27, 39, 48, 49, 53, 55, 89, 92,
    93, 99, 102, 104, 106, 107, 111, 119, 192,
    208
Straight-way cock, 221
Sub-critical flow, 181
Submersible pumps, 212
Subsurface drainage, 174, 183
Suction, 141, 142, 147, 213
Suction pipe, 148, 213
Surface drains, 104, 174, 177, 178, 180, 182
Surface water, 2–4, 9, 19, 24, 33, 153, 165
Swimming pool, 5, 60, 65, 66
Swing check valve, 217

T
Tee, 199, 200
Throttling valve, 215, 217
Time of concentration, 175, 176
Toxin, 67, 82
Transmissivity, 156–158
Treatment, 2, 5, 16, 18, 45, 46, 52, 56, 91, 108,
    117–120, 122, 128–130, 227
Trench, 164, 165, 184, 232, 233
Troughs, 164, 225
Tube wells, 19, 168

U
Ultraviolet light, 56, 126, 128, 129, 194
Ultraviolet radiation, 129, 193
Unions, 196, 197
uPVC pipe, 193, 194
Utensil washing, 61

V
Valve, 56, 145, 192, 214, 215, 218, 219, 221
Vapour, 2, 24
Velocity, 24, 120–122, 125, 141, 143, 144,
    147–149, 176, 181, 182
Vibrio, 39, 127

W
Waste cock, 221
Water, 1–3, 5, 7–9, 11, 13, 15, 19, 24, 30, 46,
    47, 54, 59–63, 65, 67–71, 89, 92, 99, 100,
    102, 106–108, 110, 112, 118, 124, 126,
    127, 129, 131, 135, 136, 138, 141,
    143–149, 153, 157, 159, 164, 168, 169,
    178, 182, 192, 205, 218, 221, 223, 224,
    227–229
Water supply fixture units, 145
Water test, 205

Y
Yield strength, 228

Printed in the United States
By Bookmasters